New Harvest

New Harvest: *Transplanting Body Parts and Reaping the Benefits,* edited by *C. Don Keyes,* 1991

Ethics and Aging, edited by *Nancy S. Jecker,* 1991

Beyond Baby M: *Ethical Issues in New Reproductive Techniques,* edited by *Dianne M. Bartels, Reinhard Priester, Dorothy E. Vawter, and Arthur L. Caplan,* 1990

Clinical Ethics: *Theory and Practice,* edited by *Barry Hoffmaster, Benjamin Freedman, and Gwen Fraser,* 1989

What Is a Person?, edited by *Michael F. Goodman,* 1988

Advocacy in Health Care, edited by *Joan H. Marks,* 1986

Which Babies Shall Live?, edited by *Thomas H. Murray and Arthur L. Caplan,* 1985

Feeling Good and Doing Better, edited by *Thomas H. Murray, Willard Gaylin, and Ruth Macklin,* 1984

Ethics and Animals, edited by *Harlan B. Miller and William H. Williams,* 1983

Profits and Professions, edited by *Wade L. Robison, Michael S. Pritchard, and Joseph Ellin,* 1983

Visions of Women, edited by *Linda A. Bell,* 1983

Medical Genetics Casebook, by *Colleen Clements,* 1982

Who Decides?, edited by *Nora K. Bell,* 1982

The Custom-Made Child?, edited by *Helen B. Holmes, Betty B. Hoskins, and Michael Gross,* 1981

Birth Contol and Controlling Birth, edited by *Helen B. Holmes, Betty B. Hoskins, and Michael Gross,* 1980

Medical Responsibility, edited by *Wade L. Robison and Michael S. Pritchard,* 1979

Contemporary Issues in Biomedical Ethics, edited by *John W. Davis, Barry Hoffmaster, and Sarah Shorten,* 1979

New Harvest

Transplanting Body Parts and Reaping the Benefits

Edited by

C. Don Keyes

Department of Philosophy, Duquesne University,
Pittsburgh, Pennsylvania

in Collaboration with Coeditor

Walter E. Wiest

Professor Emeritus, Ethics, Pittsburgh Theological Seminary,
Pittsburgh, Pennsylvania

Humana Press • Clifton, New Jersey

Library of Congress Cataloging-in-Publication Data

New harvest : transplanting body parts and reaping the benefits /
    edited by C. Don Keyes in collaboration with Walter E. Wiest.
        p. cm. — (Contemporary issues in biomedicine, ethics, and
    society).
    Includes bibliographical references (p. 281) and index.
    ISBN 0-89603-200-0
    1. Transplantation of organs, tissues, etc.—Moral and
ethical aspects. 2. Homografts—Moral and ethical aspects.
I. Keyes, C. D. (Charles Don) II. Wiest, Walter E., 1920–   .
III. Series.
RD120.7.N49 1991
174' .25—dc20   90-26049
                                                                      CIP

Preface

New Harvest includes contributions from specialists in medical, philosophical, psychological, religious, and legal fields. These essays are not simply a collection, but were developed from a single conception of the four ethical concerns of transplantation described in the first chapter. The individual chapters are all parts of a structure unified by the search for ethical foundations basic to the four concerns.

Transplantation is surrounded by a great deal of understandable emotional sensitivity. The authors trust that words like "procurement," "harvest," and possibly other expressions found in this book will not offend. We use the current language, but do so with objectivity and respect for those who are personally involved in transplantation. We have made room for, and indeed have invited, different and sometimes conflicting points of view on the complicated ethical questions raised by transplant operations. We can not assume that there is one right answer to these questions, at least at our present level of scientific knowledge and ethical wisdom. We do not presume to have identified and analyzed all the ethical questions raised with equal thoroughness. There are four ways in which the scope of the book is limited. Identifying these limitations also helps designate what it is in its own right.

First, some questions have been given more attention than others. It may well be that we have omitted some important issues altogether, but we hope there have been no such glaring omissions. In any case, we want only to offer these essays as a contribution to a continuing process of medical innovation and ethical response.

Second, the ethical perspectives from which we interpret the issues are admittedly limited to western ways of thinking. Even though they are stated in terms of western philosophy, the basic foundations of biomedical ethics have universal scope (the second chapter and elsewhere). However, other perspectives (e.g., religious ethics) are largely western, whereas the laws discussed are almost entirely those of the US.

Third, this book is limited to ethical issues related to allografts, namely transplantation from one human being into another human being. It touches upon xenografts (heterografts), the grafting of animal parts into human beings. Except in passing, it does not deal with autografts, that is, transplantation from one part of the same person's body to another part. However, it discusses more than solid organ transplantation alone, since it also includes other types of tissue grafting, cellular transfers, sperm and oocyte donation, as well as subcellular implants and biotechnology. The book focuses primarily on those types of transplantation that are done or seriously contemplated at present and secondarily and more briefly upon types that are speculative at this time, but that may be undertaken in the foreseeable future.

Fourth, much that is said applies only to the present and foreseeable future. In the more distant future, techniques of organ transplantation that now exist may well be replaced by artificial (nonorganic) devices and the use of animal organs. Recognizing that such changes might occur, nevertheless this book tries to capture and interpret some of the main ethical issues of transplantation as it exists at this particular time in history.

C. Don Keyes

Foreword

The past few years have seen tremendous advances in the field of organ transplantation to the point where transplantation today is considered the accepted treatment for end-stage organ diseases of many types. The current success in organ transplantation is attributable to advances in laboratory investigation, a better understanding of the biology of transplantation, refinements and standardization of surgical techniques, and major advances in patient care. Now that transplantation has become relatively commonplace, and major hurdles in the biology and technical aspects of this specialty have been overcome, a major new field of inquiry and importance has emerged. That is, today, the physician, the scientist, and the public, who come in contact with any aspect of transplantation are faced with a variety of psychosocial, ethical, religious, and legal issues. In no other field of medicine have such issues had so many implications and played such an important role as in the field of organ transplantation. Such a focus on transplantation has lagged behind significantly, compared to the rapid pace of the medical advances.

C. Don Keyes has enlisted the contributions of an impressive list of experts and has compiled a novel volume that admirably addresses these very important issues. *New Harvest: Transplanting Body Parts and Reaping the Benefits* takes a major leap forward in emphasizing this important area of transplantation. The book covers a spectrum of issues, and identifies major areas of concern and controversy in as constantly evolving and changing a field of medicine as transplantation is.

The book covers some very practical issues in organ transplantation, including chapters on the medical considerations of procurement and on sexual performance before and after organ transplantation. Such chapters are complemented by numerous chapters that cover much more difficult and sublime areas of consideration, including religious, ethical, and psychological concerns. There is also an excellent review of some of the legal issues germane to organ transplantation.

Dr. Keyes must be commended not only for his role as editor, but also for his numerous important contributions to the book. He has formulated what will go a long way toward bridging the gap that exists between the purely medical and scientific issues of transplantation and those ancillary issues, many of which are herein described, and which are so important for the unique nature and for the future development of transplantation. Dr. Keyes and his contributors deserve congratulations for creating a volume that will become a valuable source of reading and departure point for any individual who is involved in any way with the specialty of transplantation. The disciplines of religion, law, bioethics, and psychology will also look to this work as a reference source as they further subspecialize into and explore the field of transplantation.

Leonard Makowka

Contents

Contributors

Todd L. Demmy • *Division of Cardiothoractic Surgery, Allegheny General Hospital, Pittsburgh, Pennsylvania*

Michael Gold • *Beth El Congregation of South Hills, Pittsburgh, Pennsylvania*

Lawrence G. Hunsicker • *Department of Nephrology, University of Iowa Hospitals, Iowa City, Iowa*

Walter Jacob • *Rodef Shalom Congregation, Pittsburgh, Pennsylvania*

George and Margaret Joseph • *Pittsburgh, Pennsylvania*

David Kelly • *Department of Theology, Duquesne University, Pittsburgh, Pennsylvania*

Edward E. Kern • *Department of Psychiatry, Allegheny General Hospital, and The Medical College of Pennsylvania, Pittsburgh, Pennsylvania*

C. Don Keyes • *Department of Philosophy, Duquesne University, Pittsburgh, Pennsylvania*

George A. Magovern, Jr. • *Division of Cardiothoracic Surgery, Allegheny General Hospital, and The Medical College of Pennsylvania, Pittsburgh, Pennsylvania*

Ivan M. Naumov • *Center for Biotechnology and Bioengineering, University of Pittsburgh, Pittsburgh, Pennsylvania*

Eric C. Sutton • *Attorney, San Antonio, Texas*

Ralph E. Tarter • *School of Medicine, University of Pittsburgh, Pittsburgh, Pennsylvania*

Abraham Twerski • *Gateway Rehabilitation Center, Pittsburgh, Pennsylvania*

David H. Van Thiel • *School of Medicine, University of Pittsburgh, Pittsburgh, Pennsylvania*

Walter E. Wiest • *Pittsburgh Theological Seminary, Pittsburgh, Pennsylvania*

George L. W. Werner • *Committee on Oversight of Organ Transplantation, and Trinity Episcopal Cathedral, Pittsburgh, Pennsylvania*

James E. Wilberger, Jr. • *Department of Neurosurgery, Allegheny General Hospital, and The Medical College of Pennsylvania, Pittsburgh, Pennsylvania, and West Virginia University School of Medicine*

Joyce Willig • *Fairfield, Connecticut*

Acknowledgments

The editor thanks the many physicians, scientists, ethicists, and others who offered information and interpretations that led to the writing of *New Harvest*. Some of these are cited in the text and footnotes. A source of insight, identified there simply as "one interpreter," was indispensable. Most of the Contributors assisted the editor in ways that went beyond the writing of their own chapters and articles. Walter Wiest and Rebecca Demas deserve special thanks for helping the editor's chapters and articles to develop by criticizing early versions of the manuscript. Mohammad Azadpur and Daniel Morrison made similar contributions. They also helped with library research, as did Terry Kasely, Walter Lesch, James Quick, and William Welton. Leslie Miller and Mohammad Azadpur assisted in compiling the references. Mary Schellhammer and Cathy Weisinger typed the manuscript and suggested stylistic improvements. Other typists include Kelly Kurmin, Joan Thompson, Susan Koenig, Melissa Deardorff, and Tomi Dreibelbis.

Permission to quote material in "Body and Self-Identity" has been granted from the following sources:

Willard Gaylin, "Harvesting the Dead," *Harper's Magazine*, **249**: 1492, 23–30 (September 1974). Quoted by permission of the author and publisher.

Hans Jonas, "Against the Stream: Comments on the Definitions and Redefinition of Death," *Philosophical Essays: From Ancient Creed to Technological Man* (University of Chicago Press, 1980). Quoted by permission of the author and the publisher. © 1974 by Hans Jonas.

Stuart J. Younger et al., "Psychosocial and Ethical Implications of Organ Retrieval," *New England Journal of Medicine*, **313,** 321–323 (1985). Quoted by permission of the publisher.

PART ONE

Medical and Ethical Foundations

Four Ethical Concerns

C. Don Keyes

Introduction

Every human being is at least indirectly implicated in the promise of prolonged life. Medical technologies have pushed back death's boundaries. One result has been the redefinition of death in terms of brain function. Conversely, human life now seems to be increasingly identified with brain function and persons seem to be cores of cerebral activity attached to increasingly interchangeable body parts. Organs cease to be fixed properties owned by selves, but have become components that ought to be shared with other selves. At the same time, the economics and politics of existing transplant surgery are causing a crisis that will radically change the way health care is administered in the US. Transplantation also presents all of the traditional problems of biomedical ethics, but causes them and the principles that give rise to them to be more intertwined, sometimes contradictory to one another, and more quickly forced to their limits.

Types of Concern

Conflict of Biomedical Ethical Values

The first type of concern is the conflict of the basic ethical principles, all of which have political dimensions, precipitated by transplantation.

From: *New Harvest* Ed.: C. D. Keyes
©1991 The Humana Press Inc., Clifton, NJ

Donors and Recipients

Transplant recipients wrestle with the conflict between the desire to live and the desire not to suffer. They are confronted with questions such as whether or not delaying death is worth the cost. The cost of transplantation is more than economic. It includes "suffering years of pain, the struggle with hope and loss of hope, progress and setbacks," as one interpreter* puts it. Marital problems and sexual dysfunction occur frequently. Physical accommodation of the new organ is often difficult and medication used to prevent its rejection causes complications such as pneumonia, bleeding ulcers, gout, muscular weakness, constipation, and warts. Psychological difficulties may also arise, since emotional integration of the graft, entirely apart from the effect of medication, is sometimes traumatic.

Transplantation can also cause financial adversity. The economics of liver transplantation, for example, can drive even the affluent and their families into poverty. More is involved than the expense of surgery and hospitalization. Financially distressed recipients sometimes return to work on reduced salary, must stop working and collect welfare, undergo discrimination, or have trouble finding a job.

Potential donors struggle with ethical questions about their own bodies when they try to decide whether to sign a living will allowing their organs to save lives after they are dead. Families of a dying or newly dead loved one sometimes struggle with such uncertainties as whether donation would respect the wishes of their loved one, whether their consent in the case of the dying might prompt the staff to retrieve the organs prematurely, and so on. Prospective living related donors must undergo major surgery if they donate and possibly guilt and family pressure if they don't. A related ethical question is whether living donors ought to receive some tangible benefit for their sacrifice.

*This information and other statements attributed to "one interpreter" elsewhere in the book were obtained through private conversations.

Physician, Patient, and Community

Conflicts also affect physicians in ways that are peculiar to their professional relationship and obligations.

First, the perspectives of physician and patient will inevitably differ somewhat, even in the best of relationships. Can the desire to preserve life and the desire to reduce suffering be reconciled? Both sanctity of life and reduction of suffering aim at positive good but they conflict in some actual situations. The physician may well be preoccupied with saving life, whereas the patient may be more immediately concerned with suffering. Surgeons must balance an inclination to preserve life at all costs by considerations such as informed consent or refusal of surgery, the undesirability of sustaining life artificially when there is no hope of recovery, and the patient's right to a kind of survival that allows meaningful interaction with the world. The result is ethical struggle, conflicts of conscience.

Second, physicians potentially conflict with the community. Their primary commitment is to particular patients, but there are also broader ethical responsibilities. For instance, there are the claims of other physicians' patients who also need a scarce transplantable organ. What system of priorities is just or fair? What voice should the surrounding community have in fashioning and adopting such a system? Further, what principles and criteria should pertain generally in regard to

1. Patient's rights;
2. Standards of treatment and health care delivery; and
3. The use of the community's resources for health care as compared to other needs and for transplant operations in relation to other kinds of medical treatment?

These issues involve not only the physician's immediate community, but also the larger community of which it is a part. This leads, in turn, to consideration of questions about the role of medicine in our whole society, and of transplant operations as a significant part of it all.

Social Responsibility

The second type of concern is that the whole society has a stake in these issues, which call for the establishment of social policies. These issues cannot be answered by medical professionals and their immediate communities alone. For instance,

- What type of regulation of procurement and allocation will help reduce the gap between increasing demand for organs and their limited supply?
- How should the conflict between an overwhelming need and sparse resources be approached?
- Who pays for transplantation?
- Who is eligible for transplantation?
- Who decides which patients will receive transplants?
- Should access to medical treatment be based on need or on ability to pay?
- Is it right that some persons in need of a transplant should fail to get one because they cannot pay, while others receive one because they can?
- What is the rule of profit-making in the health care field?
- Do pharmaceutical companies charge too much for their products?
- Do hospitals and physicians charge more than is fair?
- Is it ethically wrong to profit from the misery of others?

Questions like these will be answered politically, and legislation is one way politics is expressed. Therefore, we could ask what models of the good should guide political answers. One such model is the "market metaphor," as Plough (1986) calls it. According to this approach, medical care should be put on the market and subjected to the economic functions of supply and demand because that will help assure that all of us will receive better treatment in the long run. In the US, according to Plough, the "momentum and the power now clearly reside with the advocates of market rationality." Other models are based on noneconomic concepts of the good. These are advocated by some who claim that health care is not a commodity in the ordinary sense. They aim at setting limits to what they consider excessive profit taking in transplantation.

The ethical concerns expressed by the conflict of biomedical ethical values are more basic than the distinction between individual persons and society. Respect for life must be founded, as it is in the following chapter, at a level prior to that distinction. In practical terms, the good for individual persons becomes real only when it is rooted in a healthy social order, namely a state in which politics and the legislation that comes from it are guided by the good. Aristotle states in *Nichomachean Ethics:*

> Thus it follows that the end of politics is the good for man. For even if the good is the same for the individual and the state, the good of the state clearly is the greater and more perfect thing to attain and to safeguard. The attainment of the good for one man alone is, to be sure, a source of satisfaction; yet to secure it for a nation and for states is nobler and more divine.

What Is Life?

The third and fourth types of concern arise from questions about the nature of human life. Transplant operations raise these questions in two ways: When does human life begin and end? How much can we alter a human being with our medical "tinkering" before we violate that human being's identity?

Beginning and End of Biological Life

Issues related to brain death and the status of fetuses lead to questions such as

- Is the organ (or tissue) source ethically acceptable?
- Should brain dead bodies be kept on respirators?
- Is it right to experiment on them?
- What limits should there be to experimenting on human beings?

Transplant surgery and the increasing demand for organs and tissues force us to try to answer unresolved questions about the status of embryos and brain death. Respect for the recipient's life does not justify the donor's death. Whether a potential donor is dead or not yet living must not be decided in terms of expediency for transplant surgery nor influenced by the recipient's need. Questions about when life begins and ends are not pragmatic and have to be separated from the needs that force us to ask them. Responsible answers should not be biased by the politics and economics of need. Answers should be sought through ethical reflection on scientific evidence about the biological foundations of human individuality. Furthermore, law does not answer ethical questions even though it is related to ethics. Sutton's chapters (this volume) will return to the question of law, showing how legislation affects the procurement and allocation of organs. There is an urgent practical need to clarify questions concerning when human life begins because of in vitro fertilization and experimentation with human embryos, as well as the use of fetal tissues for transplantation. Recently, procurement of fetal brain material for the treatment of Parkinson's disease has become a focal point of controversy.

Self-Identity

Theoretical concerns pertaining to brain death also raise new questions about the relation between brain and mind and, thus, about what constitutes a human being. Interpreting death as brain death seems to imply that mind and self are purely physical, or at least that a human being is the cognition which brain function produces. But are we just physical or are we more than physical? Have we "souls" as well? Historically, these questions have produced two opposing views. The first is monism, which holds that we are entirely physical or that we are entirely spiritual. The other is dualism, which holds that we are a combination of two distinct and essentially separable elements. Also, it has usually held that the soul or spirit is what we really are.

Transplantation further complicates such questions. It confronts us with the possibility that the grafting of brain tissues in the future might alter self-identity. There is also concern that genetic engineering may change what the human race will be in the future. Still another question is raised by the fact that transplantation treats the body parts of different individuals as interchangeable. Since our bodies are so much a part of our selves, transferring an organ from one person to another creates problems, sometimes severely, for both donor and recipient. Transplantation also forces us to ask questions about nature of the self. As Winslade and Ross (1986) suggest:

> Are we spirits who happen to possess bodies and in fact need those bodies in order to manifest ourselves in this particular material world? If so, then it is not much concern whether we are inhabiting a pure or mixed-parts body. Are we, instead, minds-and-bodies, a kind of computer-like system in which the bodies are our hardware and the minds our software, our operating systems, as it were? If so, then the software, like any software, can run on any compatible body, although often not as well on the body-hardware for which the mind-software was originally intended . . . individual integrity lies presumably in the brain or in the more complex parts of the nervous system culminating in the brain (although even here a serious problem lurks as researchers investigate the transportability of brain tissue). The rest, outside and inside, is mere packaging or operating parts, to be used and, when exhausted, to be replaced.

Ethical Reflection

Respect for life is the good, the "ought," on which all four issues are based. Interpreting this good in relation to transplantation, though, involves more than restating traditional values in the light of new facts, since respect for life gains new significance from the conflicts and crises caused by transplant surgery and must be reconstructed on foundations appropriate to it. All such speculation about respect for life is ethical and goes beyond science. This does not mean that ethics is, therefore, antiscientific; it lacks the verifiability of science but must always take account of scientific fact.

Ethical judgment must be distinguished from religion and law. Foundations for respect for life are reconstructed in the next chapter independently of any theistic claims. The principles built on this nonreligious base, which form the philosophical framework of this book, will be the source of ethical judgment throughout, while theologically based ethical judgments about transplantation will be discussed later in the book. Similarly, legal aspects of transplantation will be considered, but law and ethics must be distinguished. Law is the product of politics. Biomedical ethics is philosophical reflection on science and the technology of healing. Although law does not answer ethical questions, it has three indirect relations to ethics:

1. Laws often imply the ethical values that led to their enactment.
2. Ethical reflection ought to guide future legislation.
3. Since existing laws are enforced, ethical decisions must take this into account.

Ethics transcends law (we can be called on to do more than the law requires), but in making ethical decisions, we must allow for the consequences if we decide that our ethical judgments contradict the law of the land.

Deontology and Consequentialism

Ethical judgment is deontological when the good that ought to guide actions is summarized as principles that must be followed without exception and can be stated ahead of time. Duty consists in acting according to rules more than asking what effects an action is likely to produce. Ethical judgment is consequentialist, however, when it asks whether the effects of a particular action are likely to produce beneficial states of being. There are at least two consequentialist ways of evaluating possible effects, both of which are found in biomedical ethics. One focuses on the intended and reasonably foreseeable effects of a new medical or surgical procedure. The other, the "slippery slope" concern, asks about possible abuse. An example of this is questioning whether gene therapy to prevent disease might eventually be misused to alter human genetic structure in harmful ways.

The distinction between consequentialism and deontology has perhaps been overstated, as if they were irreconcilably different methods. It seems that both perspectives have more in common than is generally supposed and differ more in emphasis than in kind. Deontologists are at least implicitly concerned about the effects of acting in a particular way at some point in formulating principles. Similarly, consequentialists are at least implicitly concerned about principles when they judge that certain states of being are beneficial and others harmful.

Some current consequentialist approaches to ethics seem to be unaware of their own deontological implications. An essay by Moore (1988) is a possible example. He defends a probabilistic calculation of possible effects and also notes that professing belief in the Hippocratic Principle does not always lead to beneficial consequences.

> The ancient dogma of "first, do no harm" appears simple and obvious. But on closer examination, it seems to have been mere window dressing for the necessary hurt that has accompanied much of medical practice over the centuries. Many drugs had bad side effects; operations were disfiguring, extremely painful, and often lethal...
>
> All of these noxious traditions harmed the patient, first and foremost. The extended concept of hurting others to help

the patient possibly had its inception with Cesarean section; but with the advent of transfusion after World War I, it became widely and socially accepted. Thus, primum non nocere, which originally meant "first do no harm to the patient" was probably never true or at least never a reflection of the reality of medical practice.

With transplantation of organs from living donors, Moore claims there is now the problem of damaging persons other than the patient, although he writes that,

...most of the limits on harm to living donors are analyzed on the basis of the likelihood of success. What if the donor kidney is not quite perfect? Maybe there is a little beginning nephrosclerosis or proteinuria and some inability to concentrate the urine. While this minimal kidney disease would hardly seem to be serious for the donor, what is the impact when the kidney arrives even slightly ischemic in the recipient? What is the probability of success? The likelihood of failure?

Moore argues, however, that,

...the solution of each of these ethical questions (the kidney fitness, the donor fitness, the recipient fitness) rests on a probabilistic or statistical analysis. What is the likelihood of harm being done, as opposed to the likelihood of good accruing to the patient?

We must recognize, as Moore does, that consequences have to be taken into account and that, sometimes, this can justify the suspension of a deontological principal. This does not mean, however, that the principle has been revoked. In fact, deontological principles can, in the exceptional case, come into conflict with one another. For example, pain killers given to dying patients can shorten their lives, thus putting two deontological principles (reduction of suffering and preservation of life) into conflict. Furthermore, just because a deontological principle is not followed faithfully does not necessarily mean the principle is wrong. One deontological principle sometimes must be given priority over the other, but this does not mean that either is revoked. Consideration of the consequences decided upon is one of the bases on which we decide which principle should be given priority. At the same time, decisions based on consequences must not try to reap beneficial effects at the expense of a foundational deontological principle. A living donor, for example, must not be killed so that his organs can benefit the dying. As a result of these and other connections between deontology and consequentialism, the two approaches seem essentially involved in one another.

Common Foundation of Consequentialism and Deontology

Respect for life seems to be equally deontological and consequentialist, and forces us to recognize the elasticity of the distinction. The ethical reflection on specific issues that follows will use the distinction with this elasticity in mind, will point out conflicts between deontological and consequentialist perspectives, and also identify points at which they converge. The following chapter attempts to establish foundations of respect for life at a level more basic than the distinction between deontology and consequentialism because it will seek a single principle which has not been differentiated into act and effects.

Transplantation and Foundational Biomedical Ethical Values

C. Don Keyes

Introduction

Respect for life is the foundation for biomedical ethics. This branch of ethics differs from some others because of the transparent and self-evident nature of the good that it ought to seek. The tradition of Hippocrates expresses this in terms of doing what is beneficial to life and avoiding harm; yet medicine and surgery change lives and alter natural processes. This forces us to ask:

- What is life?
- Is it simply something "given"?
- Do we have a right to intervene unnaturally?
- What may be the limits of our tampering with life?

Whereas unnaturalness is at least a latent concern in all types of transplantation, reproductive transplantation pushes the crisis of unnaturalness to its limits. Brain death, the neurological status of fetuses, whether anencephalics are persons, speculation about grafting brain tissues and cells, and related concerns about self-identity that arise from transplantation disturb established interpretations of life and death. The lines between life and death become questionable. At the same time, reproductive transplantation

From: *New Harvest* Ed.: C. D. Keyes
©1991 The Humana Press Inc., Clifton, NJ

and brain tissue grafting push the crisis of self-identity to its limits. Conflicts involving self-identity can also occur in all the phases of the process of donating and receiving organs.

This chapter attempts to reconstruct the foundation for what is perennially valuable in Hippocratic beneficence and avoidance of harm in a way that is adequate to the conflicts caused by transplantation, especially those related to self-identity. Respect for life will be based first on philosophical reflection without referring to religious ethics. Later chapters will examine religious perspectives on transplantation ethics as independent ways of attesting to respect for life.

First Philosophical Foundation

Life, as such, is worthy of respect, but higher animals are especially worthy because they are conscious and intelligent. It is beyond the scope of this book to explain respect for nonhuman animal life, but this question is valid since the arguments used here to ground respect for human life seem to apply in varying degrees to other animals with higher brain function. Human life is fully aware of itself and its crises, including those that affect its self-identity. It is worthy of respect because it values; it answers to ethical demands. This requires standing outside the self through self-criticism and reflection on what ought to be (might be), not just what is, and taking an interest in what the other and I can become.

The neurological and related issues that push the crisis of self-identity to its limits also help to reconstruct respect for life because they force us to decide what a human life essentially is, and therein must reside the intrinsic value that demands and evokes respect. To ask what is essential to being human is also to raise the philosophical question of how brain and mind are related to one another. The reflective activity that reconstructs foundations in and through, rather than around, the crisis of self-identity must

1. explain the resistance to asking how brain and mind are related;
2. interpret that relation in a way that
   a. eliminates dualism but allows reference to mental states, thereby establishing the first foundation of respect for life;
   b. clarifies the definition of life implied by brain death criteria;
   c. designates the type of brain function that makes life worthy of respect; and
   d. explains the context in which such brain function occurs, thereby establishing the second and third foundations.

Resistance to Brain-Mind Questions

There is extraordinary resistance to asking about the brain-mind relation, and it seems to spring from more than closed-mindedness toward "philosophical" questions as such. At least three other less obvious obstacles might stand in the way of asking this particular question.

First, much of contemporary philosophy itself, not to mention psychology, treats any version of the mind-body problem as if it were a pseudo-question. A psychologist I interviewed indicated this when I asked him the admittedly naive question of how the brain produces consciousness. He argued that consciousness is nothing other than an entity's response and adaptation to its environment and, therefore, that the thermostat on the wall and computer on the desk are, in principle, "conscious." When I pointed out that neither the computer nor the thermostat is aware that it is conscious, as he and I are, he said that I was making a "metaphysical" claim. I responded that I was really asking a "physical" question about how the brain causes consciousness. He then reformulated his original answer and talked about stimulus and response. There are legitimate reasons for my colleague's, and others', refusal to deal with the brain-mind problem. Modern science seeks to understand phenomena insofar as they are observable, predictable, and able to be manipulated. To explain or account for something by introducing a nonobservable item such as "consciousness," or some other synonym for mind, is to disrupt, or at least to diverge from, scientific inquiry. Another legitimate reason contemporary philosophy and psychology avoid the brain-mind problem is their desire not to be dualistic and not to imply that human beings are split into two separate entities: a physical body and a real but intangible mind (or soul) that interact with each other closely or from a distance.

Second, others resist asking how brain and mind are related for almost the opposite reason, as stated above. They fear that the inevitable answer will destroy their values and undermine the reality they attribute to values of ethical, religious, or personal interests. This type of resistance to asking the brain-mind question dreads losing the good in a reductionistic approach to biology.

Third, what we now know about the brain, that it is the seat of emotion as well as thought, has run ahead of established symbolism that centers emotion in the heart, even when it is conceded that thinking is done in the head. Some ancients knew that the brain is related to consciousness. That fact was not generally recognized, however, until the Renaissance, and it was only in the late nineteenth century that the brain became an object of scientific study. Even some of the most elementary principles of brain function, such as synaptic transmission, were not empirically established until the twentieth century. Quite recently, the brain has also been proven to be the place in which death occurs. Human life no longer expires through the nostrils, but is produced by or attached to, depending on one's interpretation, what happens in the head. The rapid advance of neuroscience, not to mention basic disagreements among scientists within this new discipline, adds amazement and anxiety to what the brain signifies.

Resistance to asking about the relationship between the brain and the mind for these and other reasons is strong, but the facts of experience have more authority. Despite overt and subtle attempts at denial, the fact that we are conscious and that the brain is the organ of consciousness is obvious. Common sense dictates that a blow to the head, certain chemicals, illness, and the progress from infancy through maturity to senility alter consciousness because they alter brain function. New scientific information about the brain reawakens slumbering speculative questions about the brain-mind relationship. Does the brain produce the mind? If so, can mind influence what produces it by altering brain function? Or, are minds distinct from brains and do they merely temporarily inhabit brains? One typically dismisses questions like these because they are "philosophical," but transplantation forces us to ask anyway.

Limits of Reduction:
First Foundation

Philosophical reflection to reconstruct foundations for respecting life in and through the neurological self-identity crisis begins with scientific evidence. Neuroscientific descriptions of the processes are neutral to questions of value. Therefore, scientific investigation and philosophical reflection on value do not conflict with each other because they are different perspectives. At the same time, the scientific study of the brain makes it evident that there is no mind without a brain. This fact rules out substance dualism, namely the view that mind and brain are essentially independent kinds of reality.

The first foundation of philosophical reflection on respect for life consists in establishing the validity of mental states by relativizing the importance of their causal origins. Even though consciousness is entirely the product of neurological events, that fact does not make its value less valuable. An event is more than its origin. It is also a result that can be influenced by a variety of other causes. Relativizing the importance of causal origins does not mean denying their existence, but rather attributing validity to the product. Mental states are produced by neurobiological events, and mind is a shorthand statement describing such events, not a substance. I speak of mind because I am aware of the inner life supporting my outer life. For example, I am aware of my intention to raise my arm and I do it.* Mind is not separate from the physical substrate that produces it, but the fact that mental events are neurological processes with respect to their origin neither denigrates respect for life nor invalidates ethical reflection about it, because the worth of the self arises from the content and use of such processes and, in that sense, is not entirely reducible to those processes. Respect for life is based on the potential as well as actual use of neurobiological processes.

A distinction can be made between the methodological monism of the natural sciences that focus on the physical (observable) as a matter of policy and a type of philosophical monism that eliminates reference to mental states. The former can also be called "noneliminative reduction" and the latter "eliminative reductionism."

The modern foundation of substance dualism seems to be a certain way of interpreting Descartes (1641) based on, but modifying, Platonic dualism. According to this interpretation, mind is a nonphysical thinking reality that, nevertheless, interacts with the brain. Some scientific researchers support this philosophical theory. Eccles (1977), for example, claims in a work he published with Popper that "the unity of conscious experience is provided by the self-conscious mind and not by the neural machinery of the liaison areas of the cerebral hemisphere." Popular dualism persists in expressions such as, "He passed away" and "mortal remains," suggesting that the real self (soul or mind) is a substance that leaves a dispensable body and goes somewhere else.

*This formulation of my argument was suggested in a private conversation with Karen Gervais.

Monistic reduction of a certain type is as ancient as dualism, since the oldest and most persistent philosophical theory of mind interprets it as a material reality or product. The ancient atomists held this position and claimed that all reality, including mind, is made up of physical particles. According to this position, perception is caused by atoms discharged from physical objects and entering our minds, which also consist of atoms. Whereas such empirical interpretations of knowledge are among the oldest western epistemologies, few ancients believed that mind is an activity of the brain. In the Renaissance, when the brain's relation to mental activity was finally apparent, Descartes' substance dualism and variations on it became the prevailing philosophical interpretation of the brain-mind relation. The scientific study of the brain in the late nineteenth and early twentieth centuries, however, led to the contemporary view that brain and mind are inseparable. All such material theories of mind are monistically reductive in the sense that they explain mental events as neurobiological events. Monistic reductions differ, however, in the kind of reduction involved (Wimsatt, 1976) and the degree (Hooker, 1981) to which they reduce mental events to their explanatory base.

"Eliminative reductionism," at the one extreme, considers mental states as pseudo-concepts. Gilbert Ryle's (1949) claim that talking about "mind" is a mistaken use of language is an earlier twentieth century example of such philosophical behaviorism. The current type of extreme reductionism is Patricia Churchland's (1986) eliminative materialism. This position not only eliminates substance dualism, as all monistic reductions do, but it also eliminates "property dualism," namely theories that claim mental states emerge from the brain. "Folk psychology," which attributes validity to mental states, is considered the chief source of property dualism. Churchland refers to it as "intuitive psychology" and describes it as "that rough-hewn set of concepts, generalizations, and rules of thumb we all standardly used in explaining and predicting human behavior...the outcome of beliefs, desires, perceptions, expectations, goals, sensations, and so forth." She argues that the theories of folk psychology must be replaced with the "conceptual framework of a matured neuroscience." According to this kind of eliminative reductionism, mental states have no validity.

Noneliminative reduction, at the other extreme, allows reference to mental states, as the three following instances indicate.

First, Herbert Feigl's (1960) "identity theory" holds that mental and neural events differ in connotation but denote the same physical phenomenon:

> Certain neurophysiological terms denote (refer to) the very same events that are also denoted (referred to) by certain phenomenal terms . . . we may say that neurophysiological terms and the corresponding phenomenal terms, though widely differing in sense, and hence in the modes of confirmation of statements containing them, do have identical referents.

Second, Grover Maxwell's (1978) "nonmaterialist physicalism" claims "mental entities are also genuinely physical" without deemphasizing the importance of consciousness and mental states such as "...'private exprience,' subjectivity, 'raw feels,' 'what it's like to be something,' thoughts, pains, feelings, emotions, etc., as we live through them in all of their qualitative richness."

Third, Roger Sperry's (1976) "emergent interactionism" explains mental events in terms of the relation of different levels of neurobiological processes. This is not dualistic interaction between mind and brain, but rather an interplay within "hierarchy of causal controls," all based in the physiology of the brain. The various levels have "their own dynamics and associated properties that causally determine their interactions." In other words, all neurobiological processes, including the interrelation of levels, remain physical, and yet, mental events have validity within such a physical context. Thus, Sperry claims, "As the brain process comes to be understood objectively, all mental phenomena, including the generation of values, can be treated as objective causal agents in human decision making."

Noneliminative positions such as Feigl's, Maxwell's, Sperry's, and MacLean's (1990), mentioned below, are monistic in the sense that they claim mental states are neurobiological events. Unlike eliminative reductionism, however, these and similar noneliminative reductions attribute validity to mental states and allow them to be known from the inside. They thereby leave room for the kind of ethical reflection capable of reconstructing foundations of respect for life. Instead of merely "looking downward from one's experience into the hierarchy of components that constitute that experience," such noneliminative reductions can also, in Pribram's words (1986), preserve "the unique character of mental processes and their contents."

There appears to be no scientific necessity for philosophical theories that eliminate reference to mental states. Natural scientific method, as such, is neutral both towards these and opposing philosophical constructions of its data. Science describes neurobiological processes that produce mental states. What more we make of it depends on further reflection based on, but partly going beyond, what science itself provides.

Brain Death

Legal concepts of brain death require medical criteria by which death can be verified. It cannot, however, be assumed that a criterion of death provides an adequate basis for a definition of life. Life is more than simply the opposite of death, biologically defined. Life has an indispensable biological base. This does not necessarily imply, though, that human life is nothing but its base. Such reductionism, which claims it is nothing more than its base, is not required by science and it is incorrect because it does not include the broader material and social contexts in which brain function makes it possible for a person to be present. Theoretical interpretations of life that are inspired by scientific criteria of death sometimes fail to include these broader contexts in their definition of life. As a result, interpretations of this type can imply a dualistic view of humanness even though they intend to be monistic. This happens when persons come to be interpreted as cores of neocortical (or whole brain) function, from and to which a variety of parts can be removed and attached. Later in this volume, "Body and Self-Identity" returns to the problem of the dualism that such a theory might inadvertently imply.

Brain death theories do not provide a definition of life, but rather criteria for testing whether or not the physical conditions exist that make it possible for a person to be present. Recognizing the difference between criteria and definitions guarantees the validity of Goldenring's (1985) claim, which is considered further in the next chapter: "The brain-life theory simply stated is: 'Whenever a functioning human brain is present, a human being is alive'." It is legitimate to say that brain function is the criterion for testing whether a human being is alive without implying that humanness has thereby been defined. With this qualification, Goldenring's brain-life criterion, as it might be called, addresses concerns that come from transplantation's disturbance of established interpretation of life and death.

Types of Brain Function and Humanness

Humanness is brain function in the sense that the mental events which constitute human individuality without exception are neural events. The pertinent neural events, however, are not limited to the neocortex, the latest product of the brain's evolution. According to MacLean (1990), the human brain now retains the earlier stages of its evolutionary history. The "reptilian," "paleomammalian," and "neomammalian" stages make man's brain "triune," with each structure contributing its own distinctive component to consciousness. MacLean claims these brain structures are not completely autonomous, even though they operate "somewhat independently." The three formations are inter-meshed in such a way that "the 'whole' is greater than the sum of its parts, because the exchange of information among the three brain types means that each derives a greater amount of information than if it were operating alone." The neo-cortex, including the most highly developed human type, is a product of the neomammalian stage. Human beings also retain evolutionally earlier cortical formations, such as those of the hippocampus and other structures of the "limbic system." MacLean writes that, in 1952, he "resorted to Broca's descriptive term 'limbic' and used the expression 'limbic system' when referring to the cortex of the limbic lobe and the structures of the brain stem with which it has primary connections." MacLean argues that the limbic system contributes to the identity of the self and the affec-tive component of symbols. This interpretation not only supports the claim that mental states have validity, but also that the human capacity to bestow value is based on deeper neurological roots than the neocortex alone.

Second and Third Foundations: Context of Humanness

Human life is valuable because it values; it takes an affective and cognitive interest in the good. It is capable of doing this partly on account of two characteristics that can-not be reduced to brain function alone, namely, it is both embodied and situated in an environment. Self-identity, the persistence of a particular person through time and change, occurs in this broader context.

The self is embodied and, as such, belongs to the body as a whole, not merely brain function. Its embodiment includes ductless glands, autonomic nervous system. A person is also brain function expressed through and affected by physical configuration, size, and similar qualities. Complex systems of organs make brain function possible. The heart is a fitting symbol of the self. It not only pumps blood, which supplies oxygen for the brain; it also responds to emotions that express intimacy.

Since it is embodied brain function, the self is therefore situated in an environment. As one interpreter claims, "brain function is the biological base of humanness, pointing to something other, incorporating, yet going beyond the material, biological dimension," referring to the hopes and other intangible contents that are integral to conscious, purposive activity and meaningful interaction with the world. Similarly, in a private conversation, Dr. MacLean said that, "The brain has no yardstick of its own; we are always on a basket of eels."

The real duality is not between brain and mind but between the whole self and content external to it. Neurobiological processes represent this content, but it is also their yardstick.

The Self as Interpersonal

The second foundation of respect for life consists in suspending the belief that only the self exists and the outside world does not, which is called solipsism.

> ...solipsism is not a theoretical statement as much as it is a description of the egotist's practical relation to the outside world. Power hungry egotism relativizes the other and treats it, him, or her, as a property of the self....furthermore, this practical solipsism operates apart from the capability of self-criticism and is antithetical to it. (Keyes, 1989)

Because it is obvious to common sense that the outside world is real, solipsism is clearly absurd and does not merit disproof if it is the theory that only the ego exists and factual reality does not. Dr. Gerhard Werner and I discussed similar questions, and he helped me formulate two concepts about the relation between the brain and the outside world.

First, some covertly solipsistic theories seem to arise from speculation on the methods of neuroscientific research. For instance, research strategies of cognitive psychology are methodologically solipsistic (Fodor, 1981). This method is correct to the extent that it does not study external reality, but rather the ways neurobiological processes are representations of external reality. However, extrapolating that method into a foundational theory of what the self is and insulating it from criticism inadvertently turns it into a type of theoretical solipsism. The scientist as a scientist is not a solipsist, but the opposite: an empirical observer who believes in material reality. But philosophy, including that of the cognitive psychologist, goes beyond science into the realm of speculation if it treats a method of observation as if it were a doctrine about reality.

Second, the brain is more than an organ for representing reality. It is also the organ that makes it possible for human beings to have expressive and responsive interpersonal relations with one another. I respect myself because the other respects me, and the worth the other grants me is the worth I see in myself. The self has value because of, not despite, the fact that the other is also real and because the self can enter into purposeful reciprocal activity with the other. Interpersonal activity seems to be the origin of what eliminative materialists call "folk psychology," namely mental states such as, "beliefs, desires, perceptions, expectations, goals, and so forth" (Churchland, 1986) have an interpersonal origin. Perhaps, as Dr. Werner also suggests, folk psychology is retrieved from eliminative materialism precisely by situating it in the interpersonal context from which it arises, not from considering the brains of individuals in isolation.

Folk psychology draws its affective power partly from its symbolic content. Symbols are formed through reciprocal interpersonal relations. A symbol is an association of something empirical and familiar with significance that cannot be stated directly, similar to what Max Black (1962) calls an archetype:

> A systematic repertoire of ideas...which...describes, by analogical extension, some domain to which those ideas do not immediately and literally apply.

It is impossible to explain symbols in a completely scientific way, since their analogical extensions often refer to some domain that cannot be reduced entirely to what they are from a literal point of view. Furthermore, this process of extension and reference is permeated by affective qualities, and these subjective states are integral to the domain signified. Symbols are an important component by which human life values and is valuable. The body and parts are symbolic. This is expressed in the emotional struggle that some transplant recipients experience when they try to adapt to a graft that represents its deceased donor. Body parts are more than mere objects. As essential parts of the self, they also symbolize the whole self. Transplanted organs carry with them "penumbras" of meaning with which we must somehow deal. The new surgery forces us to rethink what a person is.

Limiting the Importance of Use

The third foundation of respect for life begins with disinterest in life's utility. It is similar to the first foundation that relativizes the importance of the causal origin of mental states in order to respect their uniqueness. This is not a denial of the truth of neurobiology, but instead, an act that creates a space of attention in which to attribute importance to what mental states are in themselves, not merely the mechanism that produces them. To value is to assign intrinsic importance to something. The third foundation of respect for life is an even more specific kind of disinterest; it renounces the illusion of entitlement to use the other. Life is worthy of respect, not because it is useful for something, one's own ego (or other egos), but because it values. The disposition that grasps this does so because it shifts interest from life's utility to letting it be and takes delight in its delight. Empathy for the happiness and suffering of life springs from this delight that simultaneously suspends interest in utility and attributes reality to life external to the self. That makes it possible for us to become aware of what one interpreter calls "the bond of life, the link between ourselves and others." Recognizing that the other is neither my representation nor a property of my ego enables me to see myself in others precisely because they have independent individuality. The interchangeability of body parts forces us to recognize interpersonal reality. At the same time, the peculiar nature of altruism in the donor-recipient relation, the physician-patient relation, the role of the community, social responsibility, and the ethical imperative of not treating body parts as commodities calls pure utility into question. The same imperative provokes our thinking about respect for human life by raising questions about when it begins and when it ends. This is the subject of the next chapter. These and other ethical judgments about transplant operations are derived in large part from disinterest in the utility of others, seeing them as ends in themselves. As Kant (1785) claims, we must always treat human beings as ends in themselves: "...every rational being exists as an end in himself and not merely as a means to be arbitrarily used by this or that will."

Beginning and End of Biological Life

Ivan M. Naumov, James E. Wilberger, Jr., and C. Don Keyes

He who sees things grow from the beginning
will have the finest view of them.

Aristotle

Death is not what it used to be.

Russell Scott

Beginning of Biological Life

When does life begin? Or, let us start with a prior question: When has life begun? According to evolutionists, life on planet Earth originated 3.5–4 billion years ago with the appearance of the first prehistoric prokaryotic single-cell life form. During the vast period of time that has since passed, numerous species have formed. Each species and each species' member has been inheriting and carrying, like a baton in a never-ending relay race, an increasingly smaller and smaller part of the original DNA created within the very first single-cell form that started it all as our earliest common ancestor. Life is, therefore, a continuum that began in the warm preocean soup and still exists or "lives" within the cells of each animal or plant organism on the planet. This includes us humans, of course. A very minute

From: *New Harvest* Ed.: C. D. Keyes
©1991 The Humana Press Inc., Clifton, NJ

portion of the genetic message of the first organism is within each of us. It is there to stay and will be passed on to the next generations for as long as life exists. And, inevitably, it will continue "living," cleaving, and replicating in the cells of all past and present living beings on earth as an indication that information (genetic, in this case), along with matter and energy, is one of the three basic ingredients of the universe.

Now, let us return to the original question and rephrase it: When does the biological life of a human individual begin? Scientific, legislative, theological, and philosophical authorities offer a spectrum of answers linking them to chief developmental biological events:

1. Biological life begins with the first contact of the gametes, namely the ejaculated sperm and the ovulated oocyte.
2. Biological life begins at conception, i.e., at fertilization when the male and the female haploid genetic material fuse to form the diploid embryo.
3. Biological life begins at implantation when the embryo implants, i.e., nidates in the uterine wall and starts growing *in situ*.
4. Biological life begins with the first biochemical signs of pregnancy; with the first detection of placental hormones in the maternal serum.
5. Biological life begins with the first heartbeat.
6. Biological life begins when the embryo/fetus starts to acquire sentient capacity.
7. Biological life begins with the onset of the second trimester of pregnancy, around which time the embryo develops into a recognizable human being and is referred to as a fetus.
8. Biological life begins with the formation of the neocortex.
9. Biological life begins at birth, when the offspring is physically separated from the mother's uterus and has started breathing.

A bill proposed to the US Congress in 1985 declared that the individual exists from the moment of fertilization.

This is but one proposed societal approach aimed at relieving the continuing controversy and political battle associated with the status of the unborn and the abortion issue.* But that seems to be too easy an approach or answer to the obvious problem, as is the statement that the individual dies with the cessation of cerebral activity. Although both statements are nowadays probably acceptable to a large portion of the scientific and legislative community, they certainly do not and will not settle the matter.** The biological basis of the notion that life begins at conception is expressed by Krimmel and Foley (1977): "Viewed from the

*The historically long discussion on abortion dates back to the ancient Greeks, who first debated the rights of the embryo/fetus in relation to the time of the creation of the pregnancy. According to the well known Roman Catholic viewpoint, life begins at conception, when the embryo is being "ensouled." At that point, the embryo becomes a full-fledged human individual, i. e., a person, and hence should be given full protection as the Catholic Church ruled in 1869. In 1902, the Church also ruled that an ectopic pregnancy should be given the same protection. Nevertheless, a double-effect doctrine was applied for a diseased Fallopian tube containing a fetus. It could be removed if the intention was to remove the dead tissue and alleviate the hemorrhage, or other conditions in the oviduct relating to the disease, and not to abort the fetus. It was sinful, though, to carry out an abortion with the removal of the tube as a consequence. Many of these principles were repeated in *Humanum Vitae* of 1968, or in the *Instruction on Respect for Human Life and Its Origin and on the Dignity of Procreation* of 1987. Interestingly, the original reason for the protection of the fetus has changed from the old idea of the need for baptism, to the right to life of the unborn fetus.

**In 1986, the Council of Europe accepted that human life starts at the moment of fertilization and, afterwards, develops in a continuous pattern. But, at the same time, in the Explanatory Memorandum of the Recommendation 1046 (1986) (1) on the use of human embryos and fetuses for diagnostic, therapeutic, scientific, industrial, and commercial purposes, adopted by the Council's Parliamentary Assembly, it was stated that European legal systems have not yet defined a moment of gestatiorial development when the embryo/fetus should start enjoying the rights given to human beings (including the right to inviolability of the person).

perspective of an adult organism, the zygote (fertilized egg before the first cleavage) represents the first time there is a specific form of life present and a specific identity attributable to the life." Singer and Kuhse (1986) agree that the human embryo is not a true human being, i.e., a person. They admit that "the embryo is clearly a being of some sort" but "it can be shown that the early embryo has no intrinsic value and does not have a right to life." Further, "the early embryo is obviously not a being with the mental qualities that generally distinguish members of our species from members of other species. It is reasonable to assume that it has no more awareness than a lettuce."[!] They recognize, however, the potential of the embryo to become a full-fledged person, but feel that it should not be considered a human being with rights for life until it is capable of sentience.

Some researchers believe that human life begins with the release of the sperm and the egg from the genital tracts of the male and the female, respectively. The gametes convey within themselves the potential to create a new living being, and as soon as they approach each other, life begins. Singer and Kuhse view the embryo itself as a degree closer to being a person rather than merely a collection of sperm and unfertilized egg(s). But, again, they do not find basis for the view that this difference of degree is sufficient to give the embryo a moral status. Is it moral to destroy or discard sperm, unfertilized eggs or early embryos before they acquire the capacity to feel pain or pleasure?

It is worth mentioning that the natural probability of conception in one menstrual cycle appears to be no greater than 25–27%, as found in a variety of communities and cultures. Some estimates imply that almost 40% of women have abnormal embryos in any one particular unprotected cycle. Others suggest that up to 70% of human pregnancies actually terminate by themselves. Moral questions, though, are raised only when humans intervene in the natural processes of reproduction. Thus, there are objections to barrier contraception methods that destroy the chances of forming life, and to intrauterine contraception devices that most likely destroy embryos in the uterus by creating unfavorable conditions for their implantation. Discarding or destroying an embryo formed in vitro is punishable by a new law in Australia, and forbidden by the existing guidelines of the professional or governmental bodies of other countries. It remains to be seen what rules will be established in different communities and cultures. The Warnock Committee (1984) in the United Kingdom set a fourteen-day limit after which the embryos, if not frozen, must be transferred to the uterus. This time limit is enough to allow for life in vitro and research by noninvasive methods. Leenan (1986), while discussing the legal status of the embryo in vivo and in vitro, introduced the term *"status potentialis"* for the period before implantation, and the term *"status nascendi"* for the period after implantation.

Remembering the possibility of using fetal tissues for transplants, it has been stated that fetal tissues might be used for transplantation before the fetus acquires sentience—a capacity for sensation, which includes pain. The earliest forms of sentience seem to develop not before the sixth week of gestation, but are definitely present by 18–20 weeks. A fetus selected as a transplant donor would be exposed to noxious stimuli capable of evoking pain in the process of termination and extraction from the uterus. In the absence of adequate fetal anesthesia, the stimuli associated with procedures such as surgical or suction dismemberment, or the generalized spasm of smooth muscle in response to prostoglandin induction will inarguably be noxious. Additionally, the very specific requirements for freshness of tissue to be transplanted calls for termination techniques that appreciably minimize the likelihood of fetal death before expulsion. The optimal technique, according to transplant surgeons, would be a hysterotomy, which is a surgical opening of the uterus. This, on the other hand, might result in a live birth.*

The heart starts functioning between the twenty-first and twenty-fifth day of gestation, which some regard as the beginning of life. At various times during history, quickening, or the ability to survive outside the uterus, has been used to mark the beginning of human life. By civil law, the embryo still in the womb might be considered born when its interests must be protected, but the legal fiction will usually become legal fact only if the embryo is born alive. If it is usually born dead, it is regarded as not having existed.

According to Grobstein (1988), there are six separable aspects of individuality that arise during the life history of every human being: genetic, developmental, functional, behavioral, psychic, and social. Genetic individuality refers to hereditary uniqueness, where "hereditary" means "able to be transmitted from generation to generation" and developmental individuality refers to achievements of singleness and its consequences. Functional individuality refers to diverse activities essential to survival in the envi-

---

*The quoted Recommendation 1046 (1986) (1) of the Council of Europe calls on the governments of the member states to limit the use of human embryos and fetuses and materials and tissues to purposes including transplantation, which are strictly therapeutic and for which no other means exist, and to bring legislation into line with this principle. The use of tissue from embryos or fetuses for therapeutic purposes must be exceptional and justified in the present state of knowledge by the rareness of the illness treated, the absence of any other equally effective treatment, and an obvious advantage for the beneficiary, for example, his survival. It is necessary to insist on this exceptional character in order to avoid the use of such tissue constituting a form of pressure in favor of mass abortion. The Recommendation also states that it is not permissible to keep nonviable embryos artificially alive for the purpose of obtaining usable tissues or organs.

ronment; behavioral individuality, to integrated activities of the whole in relation to environments; psychic individuality, to one's inner experiences accompanying behavior; and social individuality, to self-aware interactions within a community of individuals. Each of these aspects of individuality arises in the process of becoming an adult human being as they appear at different times. The question thus posed is which of these, or what combination of them, merits new status?

From a genetic point of view, individuality is established with fertilization, since the newly formed organism's genome is uniquely different from the genome of any other siblings (other than that of an identical twin). The other aspects of individuality are not completed until the end of puberty and beginning of adolescence. In our societies, full rights are not granted until 18–21 years of age.

Ethics of the Beginning

If human individuality depends upon the full presence of developmental, functional, behavioral, and psychic individuality, then embryos, children, and demented adults are not persons. If human individuality depends on genetic identity, then a person is present at conception (formation of the diploid cell) even though the other types of individuality are only potentially present.

Disagreement about when human life begins mainly stems from conflicting philosophical definitions of human life, and these lead to different judgments about the biological stage of development at which a person, a human individual, is present. There is also disagreement about the degree of self-identity required to constitute human individuality. Three theories about the beginning of human life seem worthy of special attention. These claim that a person is present:

1. With genetic individuality at conception ("at conception theory");
2. When the embryo has developed sufficiently to survive outside the womb ("viability theory"); or
3. At some time between conception and viability, when a development occurs that is designated as essential in a particular definition of human life.

The formation of the neocortex (the eighth position listed above) seems to be the most decisive biological change between conception and viability. This development establishes the potentiality for what is distinctly human. Nevertheless, the argument that follows claims, with certain qualifiers, that a human being is present when higher brain function begins. This criterion will not be acceptable to all, since the question of when human life begins is not purely scientific but depends partly on the questioner's philosophical position. There are understandable differences of opinion about which of the nine biological conditions listed above, as well as other possible hypotheses, is the basis of human individuality. Nevertheless, the "brain function hypothesis" is uniquely appropriate to the question, theoretically consistent, and makes use of a criterion that, at least in principle, can be scientifically tested. According to this hypothesis, the onset of brain function marks the beginning of human life just as the death of brain function marks its end. Before brain function begins, the fetus is potentially, not actually, human. However, the various kinds of "potentiality" are not obvious and need to be clarified. It could be argued that the fetus is worthy of respect as a potential human being even before brain function, because of its unique potentiality.

Goldenring's Argument

Lockwood (1985) and a number of researchers he cites argue that the fetus becomes human when brain function begins. Goldenring's (1985) formulation of this argument will be explained here because it also illustrates some important related issues. He claims that the "at conception" theory is defective because it mistakes potential for actual human life:

> The "at conception theory" says that a human being exists at the moment when haploid human gametes fuse into a fertilized diploid human egg. Essentially, this theory says that a single cell is, in fact, a human being.

The viability theory is defective, according to Goldenring, because the brain of the fetus begins to function long before it can exist viably outside the womb: "EEG activity has been demonstrated at eight weeks." Contrary to both of these positions, Goldenring argues, as explained in the preceding chapter, "the integration of the brain as a whole" ought to mark both the beginning and end of life:

> Before the brain begins to function, what we have is a set of tissues or a series of organ systems. They may be functional, but without the presence of a human brain they are unco-ordinated and without the ability to develop personality. On the other side of the curve of life, when the brain ceases its activity, what remains are unco-ordinated organ systems which may be functional. They will eventually degenerate to the tissue level and then to the level of isolated living human cells until the last cell ceases to function.

The preceding chapter claims that Goldenring's concept of the "integration of the brain as a whole" should be called a "criterion" rather than "theory" in order not to confuse a test to determine whether life is present with a definition of life. This and other issues arising from Goldenring's theory are especially relevant to the ethics of brain death later in the present chapter. Two other issues related to Goldenring's interpretation of the beginning of life are immediately important to the question of the beginning of life and potentiality.

Kinds of Potentiality

The word "potentiality" often seems to be used equivocally when referring to the beginning of human life. At least two sets of distinctions must be made concerning potentiality in order to evaluate Goldenring's position. Different types of potentiality appear to be distinguished by new levels of actualization.

Preparatory and Expressive Potentiality

There is a process of development leading to a decisive event in which a new quality is produced. This process is preparatory potentiality. By contrast, expressive potentiality is intrinsic to the event itself. It is the new quality at work, namely, becoming actual, the process by which its actuality develops in varying degrees. Both types of potentiality apply to human individuality. Prior to brain function, fetuses are potentially human only in the preparatory sense because they do not yet have actual self-identity. After the brain begins to function as a whole, the fetus is potentially human in the expressive sense, and so are the child and socially developing adult. Humanness is this second kind of potentiality, a fluid process whose development starts with brain function, progresses, regresses, and continues as long as brain function continues. It is humanness unfolding by means of the organon of brain function. If that organon exists, the unfolding takes place as expressive potentiality. Distinguishing expressive potentiality from other kinds answers an objection that could be made against the neurological hypothesis. This objection might interpret the claim that humanness is more than brain function alone to mean that brain function cannot be the criterion of the beginning of life. Since fetal brain function only potentially has functional, behavioral, and psychic individuality, it is not the kind of brain function that can constitute the presence of a person, according to this objection. Stated in the language of the preceding chapter, human identity is more than brain function alone because it is also embodied, expressive, responsive, situated, and so on. According to this objection, fetal brain function has not yet acquired that broader context by virtue of which it can be distinctively human. The objection is defective, however, because it confuses types of potentiality and thereby mistakes developed self-identity for the process of being human. Fetuses whose brains have begun to function are human beings because they are in the process of being human and must be respected, although their individuality is not yet complete.

Conditional and Unconditional Potentiality

A different kind of process leads to a decisive event and prepares for it, for example, if it has already started and if the emergence of the decisive event is its inherent outcome. By contrast, potentiality is only conditional before this process starts. Since the emergence of brain function is inherent in the process of fetal development, a fetus prior to brain function is potentially human in an unconditional sense. By contrast, sperm and oocytes are potentially human in only a conditional sense. They are merely haploid and not in a preparatory process. Distinguishing the purely conditional potentiality of sperm and oocytes from the kind of potentiality that belongs to processes reveals a possible defect in a facet of Goldenring's theory, which is not essential to his foundational argument. His claim that every human cell is a potential new human being needs to be examined critically:

> Since every single human cell contains the full genetic information of every other cell, it will become possible to take a skin cell and cause it to divide and differentiate into any desired organ, or even into a complete human being, genetically identical to the donor. The process is called "cloning." Will we then say that every skin cell is a human being? Rather should we not say that both zygotes and skin cells represent potential human beings with differing probabilities of actually becoming a full human?

Although it is true that every cell is a potential human being in some sense, only diploid fetal cells in the preparatory process of acquiring brain function are potentially human in the unconditional sense. Even though each cell of the body has genetic individuality, it is potentially human in only a conditional way. Whether a cell achieves unconditional, diploid status through natural reproduction or cloning is not the issue. Even though fetuses prior to brain function are only potentially human, the fact that they are unconditionally so means that they are worthy of an appropriate kind of respect. They are in a process that will lead to brain function if allowed to develop. Just as brain function is the criterion of humanness that marks the real beginning of human life, an organism that will develop brain function symbolizes what it already has in an unconditionally potential way. Fetuses prior to brain function are in an "intermediate condition" and are worthy of respect in some sense, even if this is not the same kind of respect they will deserve when brain function begins. Later in this volume, "Reproduction and Transplantation" returns to this question.

Current Medical Concepts of Brain Death

Within the past two decades, the medical and legal aspects of the definition and determination of death have come under great scrutiny. Abundant evidence has documented that severe brain injury or damage can completely and irreversibly destroy this organ's function and capacity to recover, even when other body systems continue to function. Thus, the concept of brain death has evolved. The American Medical Association, the National Conference and Commissioners on Uniform State Laws, the American Bar Association, the American Academy of Pediatrics Task Force on Brain Death, and the President's Commission for the Study of Ethical Problems in Medicine have all endorsed the following definition of death:

An individual who has sustained either (1) irreversible cessation of circulatory and respiratory functions or (2) irreversible cessation of all functions of the entire brain including ths brainstem is dead. A determination of death must be made in accordance with accepted medical standards.

Brain death is now universally medically accepted as being synonymous with death of the person and is covered by statute in 44 states. Three medical considerations make the concept of brain death important:

1. The ability of current medical expertise and technology to prolong existence;
2. Limited and expensive critical care resources; and
3. Organ procurement for the purposes of transplantation.

Historical Perspectives

Primitive peoples did not regard death as a natural phenomenon, since they rarely survived to an old age. From mythical and religious stand points, it was assumed that as the soul left the body, respiration ceased. Thus, early physicians came to rely on cessation of respiration to establish death. After the circulation was discovered and ausculation was introduced in 1819, the absence of heart action become the accepted method of determining death. However, it was often difficult or impossible to be sure that respiration or circulation had stopped. This led to erroneous pronouncements of death and, occasionally, to premature burials. Because of this, numerous schemes and devices were conceived to ensure that resuscitation did not occur or, if it did, that the deceased would be rescued before burial. As Marshall (1967) reports, Colquet and Laborde suggested placing a bright steel needle into a muscle. Lack of tarnishing on its removal was considered a definite sign of death. Middledorf also suggested the use of a needle, but his was four inches long and bore a small flag. It was to be thrust into the heart so that the flag would wave if the heart was beating.

In the last half of the nineteenth century, the fear of premature burial reached hysterical proportions in France. The Paris Academy of Sciences awarded prizes on the subject of preventing premature burial, according to Marshall. The prize of 1890 went to a Dr. Maze, who considered putrefaction the only definite sign of death and advocated the provision of mortuaries in cemeteries so that bodies could lie there until decomposition became evident. In 1900, the prize went to Dr. Icard of Marseilles, who recommended the injection of flourescein beneath the skin. If the person were living, the whole skin would turn yellow and the eyes green within a few minutes.

By the mid-twentieth century, absolute determination of cessation of cardiorespiratory function no longer raised concerns. On the contrary, ethical debates arose over the technological advances that allowed prolongation of cardiac and pulmonary function in spite of signs of absence of brain function. This issue was first addressed in 1959 by Molleret and Goulon, who introduced the term "coma dépassé"— beyond coma—to define brain injury/damage felt to be irreversible. The first recognized criteria for the determination of brain death were provided by the Harvard Ad Hoc Committee on Brain Death in 1968. Since then, many medical and legal groups have formulated criteria to assist physicians in the determination of brain death. In the intervening two decades brain death has come to be widely accepted in the medical and legal communities and, to a significant although lesser extent, in the public mind. Public acceptance of the concept of brain death has steadily increased. A survey reported in the *Journal of the American Medical Association* in 1985 (Manninen and Evans) indicated that while 75% of those polled were aware of brain death, only 55% approved of its use as the legal definition of death. A similar survey in 1987 (Veith et al.) found a 79% acceptance of brain death as legal death.

Physiology of Brain Death

The declaration of brain death is based on the premise of complete and irreversible loss of all higher cortical (neocortex) function as well as cessation of all brainstem activity. The neocortex is responsible for maintaining awareness of and responsivity to the external environment. Coma develops when neocortical function is suppressed or ceases. Thus, loss of neocortical function results in coma with no eye opening, no spontaneous movement, and no movement produced by painful stimulation of the head and/or trunk. The brainstem controls the regulation of respiration as well as certain basic protective reflexes:

1. Pupillary reflex—a strong light directed into either eye should result in a decrease in the size of the pupils in both eyes. Any pupillary reaction indicates brainstem function and excludes the diagnosis of brain death.
2. Extraocular (eye) movements—as the head is rotated from side to side, the eyes of an individual in coma will rotate spontaneously in the opposite direction (oculocephalic or Doll's eye reflex). If this reflex is absent, a more potent stimulus is administered in the form of ice water irrigation of the ear canals. This stimulus should result in rapid movement of the eyes away from the stimulus (oculovestibular reflex). Any movement of the eyes indicates brainstem function.
3. Corneal reflex—a light touching of the cornea results in blinking of both eyes. Any movement of the eyelids suggests ongoing brainstem function.
4. Gag reflex—touching any obiect to the back of the throat results in gagging. Any movement of the palate or retching indicates brainstem function.
5. Cough reflex—stimulation of the upper part of the airway or trachea results in coughing. Any sign of movement or coughing with this stimulation excludes the diagnosis of brain death.

Each of the above reflexes requires intricate coordination by nerve centers and connections within the brainstem. The presence of any of these reflexes, however minimal, indicates ongoing brainstem activity and precludes the diagnosis of brain death.

Respiration is driven by several centers in the brainstem, which are in part dependent on blood oxygen (PO_2) and carbon dioxide level (PCO_2) levels. The lower the PO_2 and the higher the PCO_2, the greater the stimulus to breathing, assuming the brainstem centers are intact. Normal PO_2 levels are in the range of 90–100, whereas normal PCO_2 levels are in the range of 30–40. The strongest stimulus to breathing is an elevated PCO_2 level. If the PCO_2 is below 30, there is little stimulus to breathe and apnea may result. Similarly, if the brainstem centers are nonfunctional, apnea results. Apnea, in the presence of elevated PCO_2 levels or significantly diminished PO_2 levels, indicates absence of function of the brainstem.

There are, however, certain situations in which neocortical and brainstem function are potentially reversibly suppressed to the point where brain death appears to be present.

Alcohol, barbiturates, and other sedative-depressant drugs, if present in sufficiently high levels in the blood, may suppress neocortical and brainstem function sufficiently to mimic brain death. Certain metabolic derangements of the liver or kidneys may also reversibly impair neurologic functions. A high level of suspicion must be maintained for the possible involvement of drugs and/or alcohol as many individuals involved in automobile accidents may be under the influence of these substances. Blood and/or urine toxocologic screening may not pick up all possible drugs potentially causing coma. It is for this reason that all brain death criteria included a mandatory waiting period of, usually, several hours between the initial examination indicating brain death and the final certification of brain death. This allows for the influence of any drugs that might be present and impacting neurologic function to dissipate.

Both hypothermia and severe hypotension may also simulate the clinical picture of brain death. If body core temperature drops below 32°C or 90°F, brain function may be reversibly lost. Similarly, a systolic blood pressure of less than 90 mm may result in neurologic findings consistent with brain death. Thus, an absolute determination of brain death cannot be established unless body temperature is near normal and blood pressure is at acceptable levels.

Determination of Brain Death

To be declared brain dead, a patient must meet certain clinical criteria. Whereas there are no universally accepted criteria, there is general agreement regarding the main clinical principles on which brain death is established:

1. Absence of cerebral reaction to any form of external stimulation;
2. Absence of respiratory activity or apnea; and
3. Absence of brainstem reflexes.

In addition to the clinical criteria, several ancillary tests have been advocated to confirm brain death. The standard confirmatory test for brain death has traditionally been the electroencephalogram (EEG), which measures neocortical electrical activity. A "flat" or isoelectric EEG, when obtained under certain rigid EEG criteria, is indicative of electrocerebral silence or brain death. A more sensitive determinant of brain death, however, is the establishment of lack of blood flow to the brain. This may be established directly by the use of angiography—the injection of dye into the cerebral arteries. If no dye can be detected within the intracranial compartment on subsequent radiographs, then confirmation of lack of blood flow to the brain is established. No organ can maintain or recover function when there is no blood flow to that organ. Other tests aimed at documenting the absence of cerebral blood flow include xenoncerebral blood flow studies and radionuclide brain imaging.

In general, each university medical center or community hospital develop their own set of brain death criteria within the scope of the above general guidelines and existing legislature. The following represents the current clinical practice for the determination of brain death at the Allegheny General Hospital, a level I Trauma Center, in Pittsburgh, Pennsylvania:

1. A six hour period from admission of the patient to the hospital and initial neurologic evaluation that indicates an absence of neocortical and brainstem activity must elapse before a patient may be certified to be clinically brain dead.
2. The clinical determination of brain death must be made and properly noted on the patient's record by an attending neurologist or neurosurgeon.
3. The cause for brain death, where possible, should be ascertained (for example, brain hemorrhage, trauma, and so on).
4. Clinical criteria:
   a. Deep coma—the patient should have no spontaneous movements and have no response to noxious stimuli
   b. There must be no brainstem reflexes: the pupils are fixed without responsivity to light, no corneal reflexes, no oculocephalic or oculovestibular reflexes, no response to upper or lower airway stimulation.
   c. Apnea: with a PCO_2 at normal levels (35–40) and an adequate level of oxygen, if the patient does not breathe spontaneously within 10 minutes and/or the PCO_2 rises above 60, apnea is present.
5. Absence of complicating conditions—the absence of hypothermia (body temperature over 32.2°C) hypotension (systolic BP over 90 mm) and the absence of intoxicating levels of central nervous system depressant drugs as documented by drug levels.

6. Ancillary investigations—an electroencephalogram showing electrocerebral silence and/or absence of brain blood flow on radionucleotide brain imaging or cerebral angiography.

Special Considerations

There continues to be some controversy over the declaration of brain death in young children. This controversy exists because it is generally felt that the immature brain is more resistant to insults leading to brain death and that even severe brain insults in very young infants may be recoverable. Additionally, it is often quite difficult to accurately assess brain function clinically in the newborn. As a result of these concerns, the American Academy of Pediatrics Task Force on determination of brain death issued these guidelines in late 1987:

1. Infants, seven days to two months old—there must be two separate clinical examinations and two separate isoelectric EEG's separated by 48 hours of observation before brain death can be declared.
2. Two months to one year old—there must be two separate clinical examinations and two separate isoelectric EEG's separated by at least 24 hours of observation before brain death can be declared.
3. Greater than one year old—two separate clinical examinations separated by 12 hours of observation. Confirmation by ancillary tests is not necessary.

Another issue to gain attention recently is that of anencephaly. Anencephaly refers to a congenital maldevelopment of the brain in which all cortical structures are absent and only the brainstem remains. At birth, respiration and cardiac activity are present because of the functioning brainstem. However, these generally cease within two to three days and death ensues. This always fatal malformation occurs in six of 10,000 births. Improved prenatal screening has led to an ability to detect anencephaly early on in pregnancy. When such a condition is detected, abortion is recommended. In late 1987, a Canadian couple with an anencephalic fetus decided not to pursue abortion, but rather to deliver the child and to donate its organs for transplantation. The obvious ethical problem for the medical community was that, with brainstem activity being present, the infant could not be declared brain dead and, thus, organs could not be immediately taken for transplantation. Anencephalics have normally not been considered as organ donors because their organs have deteriorated too much by the time they are legally brain dead. Complicated ethical and legal debates thus ensued over the couple's request to donate the child's organs immediately after birth. On delivery, the infant was placed on complete life-support systems to maintain organ viability. Two days later, a three-physician panel concluded that all brain activity had ceased and declared brain death. The infant's heart was subsequently transplanted into a two and a half hour old child with severe heart disease. Another case has subsequently occurred in the US. However, the issue of determination of brain death became moot when the infant was stillborn. Certainly, continuing medical, ethical, and legal controversy will evolve over the question of anencephaly and the appropriateness of declaring brain death in infants whose parents wish to donate their organs.

Ethics of the End

Harrison (1986) argues that brain death laws should be revised to allow anencephalic organ donations with parental consent, since "[w]hole-brain definition of death was drafted to protect the comatose patient whose whole injured brain might recover function." The anencephalic infant, however, "lacks the physical structure (forebrain) necessary for characteristic human activity, and thus can never become a human 'person'." Capron (1987) objects to using anencephalic infants as organ donors because they are not brain dead by whole brain standards, and the standards implied by such donations might lead to abuses starting with vegetative state organ donation:

> Like the anencephalic babies, Ms. Quinlan and other patients in a persistent vegetative state lack the ability to think, to communicate, and probably even to process any sensations of pain and pleasure (at least in the way that we think of these phenomena).

Whether patients in the vegetative state should be considered as organ donors is potentially an even more controversial issue. Legislation to allow anencephalic organ donation expresses or implies the view that the definition of death should be revised to include cerebral death. Acceptance of this standard would allow donation from patients in an irreversibly vegetative state without the "slippery slope" danger of mistreatment of persons who might be considered subhuman on other grounds (e.g., racist ones). Vegetative-state patients are in an "irreversible noncognitive state" in which there is "bilateral destruction of the cortex" and possibly other structures (Korein, 1986). By higher brain death standards, human individuality is no longer present, but by whole brain death standards such patients are still alive since enough of the brainstem is living that they can breath, often without artificial support systems. Korein (1978), who testified in the Karen Quinlan case, argues that persons in the vegetative state are not brain dead. But are such persons human beings with a full "right to life"?

Conflicting Interpretations

Anencephaly and the vegetative state complicate the question of whether whole brain death or higher brain death criteria should be used to determine death. This question must be asked from a foundational perspective

by first contrasting the whole brain and neocortical interpretations of death.

Whole Brain Death

The Uniform Determination of Death Act, which identifies brain death as "irreversible cessation of all functions
of the entire brain, including the brainstem," is becoming
the legal norm in the US. The soundest argument in favor
of maintaining the whole brain death criterion as the legal
norm is that it protects patients with reversible comas from
mistaken diagnosis. This argument appears irrefutable for
obvious practical reasons unless higher brain death can be
distinguished from reversible coma with complete scientific
accuracy. This type of concern must be distinguished from
the objection that revision of the whole brain death criterion
might lead to even more extreme "slippery slope" abuses.
There are arguments in favor of whole brain death that
seem to go beyond concerns about mistaken diagnosis and
possible abuse.

First, some scientifically based investigations cited by
Horan (1978) have raised questions about whether the
brainstem might produce some "primitive psychic activity." If so, then a patient with a living brainstem would
have some claim to humanity even though higher brain
centers are dead.

Second, a philosophical-type argument is sometimes
based on observations like the one in the preceding chapter
that humanness is more than brain function alone because
it is also embodied, and so on. Concepts like this are then
elaborated into an argument to defend whole brain death:
Human life requires both (1) the whole body (including all
of the brain's parts) and (2) the functions that the parts of
the brain make possible (including breathing sustained
by the brainstem and consciousness, thought, and so on,
sustained by the neocortex).

Both of these arguments are subject to objections and
qualifications.

Neocortical Death

First, with respect to the argument that the brainstem might produce some primitive psychic activity, evidence suggests that this is limited to reflex. The brainstem might mechanically react to pain as an automatic response, but it appears that pain is not felt apart from the neocortex. As stated in "Physiology of Brain Death," in this chapter, the neocortex is "responsible for maintaining awareness of and responsivity to the external environment." Consciousness, thought, emotion, sensation, and the function of the brain as a whole are no longer possible when the neocortex dies. Even though other brain structures are clearly involved in emotions, memory, alertness, and so on, there is no evidence that they produce or sustain those mental functions on their own in a human way apart from the neocortex. Scientific research will provide new information about the neurological base of consciousness, and ethical reflection must change if necessary. Nevertheless, the fact that the human mind is impossible without a living neocortex seems to be established.

Second, the philosophical argument in favor of whole brain death seems to confuse a definition of human life with the criteria by which we determine its presence or absence. Human individuality must be defined in terms of more than brain function alone, since expressive potentiality works itself out in the broader context of the self as "embodied, situated in an environment," and so on. The argument in question safeguards that truth, but mistakenly assumes that the neocortical criterion of death would deny it. This assumption is incorrect since neocortical activity, even though the sine qua non of human individuality, is not a container that limits how we define what it makes possible.

Death of the Brain as a Whole

Goldenring's theory that human individuality is based on the "integration of the brain as a whole" seems to mediate between neocortical and whole brain death standards and is parallel with his criterion of the beginning of human life. This criterion states, as explained earlier, that "whenever a functioning human brain is present, a human being is alive." Since humanness depends upon the integration of the brain as a whole, Goldenring argues that brain death is "'death of the brain as a whole' and not the death of the 'whole brain' (i.e., of every single cell)..." The advantage of this position is that it provides a point of departure for two additional arguments.

First, Goldenring's position theoretically leaves room for contributions that other parts of the brain, besides the neocortex, make to human individuality:

> The brain can be divided into cortical and subcortical sections. The former controls most associative (i.e., thinking) structures. The latter 'primitive brain' influences behavior and emotion, but primarily is concerned with regulating body functions.

It might be possible to view cortical, not merely neocortical, function as the basic neurological condition of human individuality. According to MacLean (1990), the limbic cortex also appears to have an important role in affect and the formation of symbols, activities essential to the definition of humanness (*see* "Transplantation and Foundational Biomedical Ethical Values"). MacLean writes that "In mammals, the evolutionary old cortex is located in a large convolution that Broca called the great limbic lobe because it surrounds the brainstem." If we accept cortical death as the criterion for determining death, then the old cortex must also be dead. Practically speaking, however, cortical life is not a viable concept unless the brain can function as a whole, and it is most unlikely that the older cortex could be alive if the neocortex were dead. The blood supply to the neocortex passes through the limbic region, and an injury capable of destroying the neocortex would probably destroy the older cortex as well. Furthermore, the older cortex is more vulnerable to oxygen deprivation than the neocortex.

Second, "death of the brain as a whole" must be construed to include the death of the neocortex . Whereas human individuality is more than neocortical function alone, for reasons already stated, a living neocortex is the foundational requirement of human individuality. A partly functioning neocortex seems to indicate that a person is still actually present, even if the degree of that function is minimal.

Brain Function and Respect for Life

Human life continues to be worthy of respect, even when functional, behavioral, and psychic individuality have receded. Hence, demented persons continue to have human individuality. As in the case of fetuses whose brains have started to function, it is not the degree, but rather the existence, of expressive potentiality, that announces the presence of a person.

Embryos and fetuses, prior to brain function, are worthy of a type of respect because human self-identity belongs to their future; they are unconditionally potentially human. At the opposite end of the spectrum, organic structures that have a past of human self-identity and were once embodiments of brain function, are worthy of another type of respect. Hence, bodies in an equivocal condition between life and death, body parts, and bodies that are unquestionably dead symbolize the brain function, hence, human individuality, with which they were integrated. Violence is done to that individuality by acts such as keeping brain dead bodies "alive" for experimental purposes, commercializing brain dead bodies and their parts, failure to respect the recipient's struggle to adapt emotionally to a new organ, or disregard for a potential donor's struggle to decide. Patients in the vegetative state, whose vital functions continue even though their higher brain centers are dead, are no longer human individuals in reality. Nevertheless, they continue to be human individuals in some residual and effectively important symbolical sense because they once had functional brains. This might partly explain public reluctance to consider vegetative-state patients as organ donors. There appears to be no purely deontological principal of respect for life that prevents vegetative-state organ donation with proper consent. At the same time, there may be impelling consequential reasons, including emotional harm to the living and "slippery slope" abuses, not to do so. As a result, the question seems to require further serious ethical scrutiny that also takes this kind of consequentialist objection into account.

Anencephalics differ from vegetative-state patients because infants and fetuses in this condition never had and can never have higher brain function. They lack a forebrain, and, therefore, have neither a history of nor capability for human individuality, even though their brainstems are alive. Whereas a person was never and cannot be present, anencephalics are worthy of respect as products of human conception, and therefore, parental consent appears to be required if they are to be considered as organ donors. At the same time, the rights of parents to decline such donation and even to try to preserve the vital functions of their anencephalic infants must also be respected. With these qualifications, it seems that no further ethical principle limits considering new legislation to allow anencephalic organ donation.

PART TWO

Medical and Historical Perspectives

Medical Considerations of Procurement

Lawrence G. Hunsicker

Introduction

Any discussion of the ethics of retrieval of organs and tissues from human sources for transplantation must surely be informed by a clear understanding of the medical issues involved. On the one hand, one has to be concerned with the safety and dignity of the donor. The morality of taking organs or tissues from a living human depends on the risk, discomfort, and inconvenience to the donor, as well as attention to the authenticity of the donor's consent, whereas the morality of retrieval of cadaveric organs and tissues turns largely on the assurances that the donor is, indeed, dead. On the other hand, the motivating factor for organ and tissue donation is the ethical good to be obtained from it, and this is determined in large measure by the benefits realized by the recipient(s) of the retrieved organs and tissues. These are all issues of medical fact or judgment. Informed ethical debate must deal with these medical realities, and it is the traditional role of the physician, like the expert witness in court, to set out the scientific realities, the likely medical consequences of ethical choices, in an objective fashion with which the ethicist can deal. It should also be recognized, though, that there is not an objective and well agreed upon line between medical fact and ethical judgment. Though physicians are concerned every day with the informed

From: *New Harvest* Ed.: C. D. Keyes
©1991 The Humana Press Inc., Clifton, NJ

consent of their patients, the concept of free choice implied by informed consent—the ability of an individual to make a decision "free" of pressures such as the authority of the physician or the expectations of a family—is basically foreign to the scientific process that assumes a cause for every event. Even the concept of life itself cannot be precisely defined in biological terms. The medical ethicist must be aware that what physicians offer as medical fact or informed medical judgment may be strongly conditioned by unconscious ethical opinions. This chapter will outline the medical realities with which the ethical debate concerning organ and tissue recovery must deal, but also call attention to some of the ethical ambiguities surrounding issues generally thought to be purely medical.

The Living Donor

There are at least three major ethical issues concerning the use of living individuals as donors of organs and tissues for transplantation, and, therefore, three main medical matters that must be addressed:

1. The safety and convenience of the donor;
2. The uniqueness of the living donor as a suitable source of organs and tissues for transplantation; and
3. The extent to which the consent of the donor is authentic.

It is often stated that the safety of the donor must be protected absolutely, and that informed consent must be freely given. As a practical matter, safety and free consent have not been considered to be entirely independent of need in judging the legitimacy of using living donors. Generally, the smaller the risk to the donor and the more clear the need for the tissue to be obtained, the less the concern over the niceties of free consent.

Blood Donation

The oldest widespread form of tissue transplantation is transfusion of blood and blood components, which developed during World War II. The risks of blood donation are extremely small (though not entirely absent), and the discomfort and inconvenience are fairly trivial. Donated blood is replaced to the donor within a matter of days, so that there appear to be no long-term risks to the donor beyond that of the initial phlebotomy. There is no question of the important medical role that has been played by transfusion therapy, and the quantities of blood needed can clearly not be obtained from cadaver or nonhuman sources. That is, there are truly no medical alternatives to the use of donated human blood. It is perhaps not surprising, then, that substantial social pressure, such as nonconfidential corporate blood drives, has been used to encourage donations. Indeed, to a large extent, one's blood has been treated as a commodity, free to sell. The recent abandonment of payment for donation of whole blood was motivated more by considerations of public health than by any concern for the ethical integrity of the donation process, and even now individuals can sell certain blood components at commercial blood centers. It is a measure of how noncontroversial blood donation is that it is usually not considered in discussions of transplantation ethics.

Bone Marrow Donation

Although closely related to blood donation, the donation of bone marrow presents some quite different issues. Bone marrow transplantation has become increasingly more common over the past several years and is now being used to treat several forms of fatal blood conditions and leukemias. The evidence is quite strong that the rate of survival of individuals receiving bone marrow transplants for conditions such as aplastic anemia and myelocytic leukemia is substantially better than that of individuals treated with conventional medical therapy. The major limitation of bone marrow transplantation is that the donor of the marrow must have tissue antigens closely related to those of the recipient. Whereas red blood cells generally have to be matched for a single genetic locus (ABO blood type) of limited heterogeneity, bone marrow must be matched for at least three gene loci of extremely great heterogeneity. Even though the rarest ABO type is found in about one individual in one hundred, the chances of a perfect match with a given individual for a bone marrow type is more in the order of one in tens of thousands. Because of the close linkage of the three most relevant gene loci, a pair of siblings has about a one in four chance of being genetically identical at these loci, and the large majority of bone marrow transplants have been between siblings. Unfortunately, only about one patient in six has a well matched sibling donor and, until recently, those without such a sibling could not be considered for bone marrow transplantation. Recent advances in the management of the consequences of marrow transplantation have made it possible for successful transplants to be performed between less well matched family members or even unrelated individuals. Unlike the situation with blood components where the limited diversity permits banking of all of the common blood types, marrow banking is clearly impractical. If a bone marrow candidate lacks a matched sibling donor, it is necessary to screen computerized lists of tens of thousands of unrelated individuals to find the one who may be an acceptable donor. Furthermore, the donation must be made at a critical time in the preparation of the recipient and the donor and recipient must be in the same hospital. Thus, marrow donors do not have the same type of anonymity as blood donors. There are currently two major registries of individuals who have expressed their willingness to be marrow donors should a patient need their type of marrow.

Bone marrow, like blood, is completely replaced within a matter of days following a donation and donation carries no long-term risk to the donor. Marrow donation, however, is a much more substantial undertaking than blood donation, requiring about a dozen large needles being placed into the bones of the pelvis. The removal of the marrow is quite painful. Generally, therefore, the bone marrow removal is performed with the donor under anesthesia and is associated with a definable risk of mortality (on the order of 1:10,000). Because of the need for anesthesia and the time required, the donation generally requires about two days of hospitalization. The risk to the donor, then, is intermediate between that of blood donation and that of organ donation. This intermediate risk, together with the unique ability of certain individuals to serve as donors of a potentially life-saving tissue, have led to some uncertainty as to how marrow donation should be treated from an ethical and legal point of view. There have been moral pressure and lawsuits (though so far unsuccessful) to force well matched but unwilling relatives to be donors and to force registries and hospitals to reveal the identities of well matched unrelated individuals whose tissue type had been determined for reasons other than marrow donation. There has been no hesitancy to accept the free donations of individuals who are complete strangers to the recipient and little concern for assuring the authenticity of the consent, but there has been a general consensus that such donors should not be compensated beyond the medical costs incurred because of the donation.

Organ Donation

Unlike the donor of blood or bone marrow, the living donor of an organ cannot replace the tissue that has been donated. The safety of donation in this situation is dependent on the ability of the donor to remain healthy after removal of an organ or part of one, and on the safety of the donation surgery itself. Medical safety has been best demonstrated in the case of kidney donation. The immediate risk of donor nephrectomy in a healthy individual is basically that of anesthesia (less than 1:10,000). However, deaths have been reported after donor nephrectomies. The major long-term risk of kidney donation is of disease or injury to the remaining kidney, loss of which would lead to renal failure. Virtually all serious renal diseases other than tumor, though, are bilateral and would lead to renal failure even in patients with two kidneys. The risk of renal injury or tumor has been estimated from insurance records to be less than the risk of a fatal automobile injury of someone who drives several thousand miles a year. Recently, because of evidence that loss of kidney tissue for whatever reason might lead to progressive renal injury, considerable attention has been paid to the effects of nephrectomy on long-term renal function. There is now a substantial consensus that this risk is minimal. Thus, it seems that the risks of renal donation are not much greater than those of bone marrow donation. There is, however, a much greater degree of pain and inconvenience. The average length of hospitalization for a kidney donor is seven to ten days, and there is considerable postoperative pain. The donor may not be able to return to work for several weeks.

Unlike blood and bone marrow, kidneys for transplantation can be obtained from cadaver donors. The original justification for using living donors was that kidney transplants obtained from closely matched relatives had a much greater chance of success than kidneys from random cadaver donors. As with bone marrow transplantation, there is a one in four chance that a sibling will share with a potential recipient all of the important tissue antigens. Kidneys transplanted from such a donor have a better than 90% one-year graft success rate, and about half are still functioning up to 20 years later. Until about 10 years ago, the rate of success of kidney transplants from unrelated cadaver donors was less than 50% at the end of one year, and fewer than 25% were functioning at the end of seven years. Further, the amount of immunosuppression required for a less well matched graft is greater and the recipient's general health and rehabilitation are often less good. Over the past ten years there have been striking improvements in the management of recipients of cadaveric renal transplants. Many centers are now obtaining results with cadaver grafts as good as those with related donors. At the same time, however, these improvements in graft outcome have led to a marked increase in the number of individuals seeking kidney transplants, and the demand for cadaver kidneys has clearly outstripped the supply. This has led to the continued use of living kidney donors, but now with an entirely different ethical justification—that only by use of a living donor can a patient hope to receive a kidney transplant in a reasonable period of time. Thus, living kidney donors are now being used not because they provide a biologically unique resource, but simply because of the practical unavailability of cadaver kidney. This change in the ethical rationale for kidney donation from living donors has also led to a significant widening of the pool of individuals considered as potential donors. First, less well matched blood relatives were considered and, more recently, biologically unrelated but emotionally close individuals (e.g., spouses) have been accepted as donors.

The acceptance of increasing numbers of less closely related living kidney donors and concerns over possible abuses of paid donors have led to reexamination of the nature of the consent obtained from these donors. Some medical professionals have argued that only those donors with a life-long, strong, family relationship can be close enough to a recipient to consent authentically to the donation procedure. Individuals reasoning in this way have usually required very careful psychological screening of more distantly related or unrelated potential donors. Other professionals have pointed out that family members, particularly twins, wives, and mothers, frequently feel substantial pressure to consent to donation, and that their consent is less likely to be "free" than that of someone less closely related. These professionals consider that particular care must be taken to assure the free consent of these close potential donors. There is a lack of consensus in the medical community as to what constitutes the "free and informed consent" that we all agree we ought to obtain from the living organ donor. That this question has not been resolved reflects that it is not, in fact, a medical or scientific question, but rather an ethical and philosophical question about the nature of humankind and freedom.

The Cadaveric Donor

There is widespread agreement that one cannot harm a dead person by removing organs and tissues for transplantation and that the good obtained by individuals and society outweighs the risk of disrespect to the dead. The major ethical issues involved in the retrieval of cadaveric organs and tissues for transplantation have centered around the definition of death, the relative priority of the desires of the deceased and of the next of kin, and the benefit of society as a whole in disposing of the deceased's remains.

It is widely assumed that the definition of death is a purely medical issue. This is really not the case. In fact, as it was discovered at the end of the nineteenth century, there is no scientifically consistent way even to define life. Cells and organisms have a normal chemistry and physiology, and when this normal function is sufficiently disrupted, we may say that the cell or organism is "dead." This definition, though, does not distinguish in any precise manner between dying and death. Applied to a human being, one would certainly not want to say that someone suffering with a terminal malignancy was already dead, no matter how impossible was the return to normality. Traditionally, death of a human being has been declared when the heart stopped beating, but this moment does not identify the cessation of physiology of all parts of the body. For instance, hair and finger nails continue to grow for days after cessation of heartbeat. One can interpret death at the time of cessation of heartbeating as the end of any meaningful interaction of the individual with the surroundings and the inevitability of cessation of all normal physiology—bodily corruption. This definition was sufficiently precise throughout the history of humankind until the appearance of two medical developments within the last several decades: the ability to restart hearts that had stopped beating and the ability to prolong function of some parts of the body, such as the heart and circulation, long after loss of ability of the individual to react in any meaningful way with the environment. The first of these advances required us to redefine death as permanent cessation of the heartbeat, though the judgment of when the cessation of heartbeat has become "permanent" is not a simple thing. Our ability to maintain the heartbeat and circulation of individuals without any brain function, thus separating the two traditional components of death by periods of days to weeks, led to the proposal and adoption of the Uniform Determination of Death Acts. These laws extended the definition of death to include death by brain criteria in addition to the traditional death by cessation of heartbeat. It is important to recall that the issue of death by brain criteria was first raised in relation to the tragedy of marked prolongation of what was judged to be the meaningless existence of unconscious patients maintained on respirators, not in the context of organ or tissue retrieval for transplantation.

To be acceptable to the public, the definition of death by brain criteria had to meet two standards. It had to adhere as closely as possible to the traditional understanding of death, and it had to distinguish clearly between brain death and other less serious degrees of brain injury. There is a spectrum of brain dysfunction, in addition to brain death, including such conditions as the vegetative state, in which patients are awake but totally unresponsive to their surroundings, and coma, in which patients are unconscious but still breathe spontaneously and have other brainstem reflexes. These states differ from brain death not only in the degree of loss of brain function but also in the prognosis for recovery. In medical practice as well as in law, there is a clear distinction between the vegetative state and coma on the one hand, and death by whole brain criteria on the other.

The ability to declare death before the cessation of the circulation is critical to the practice of organ transplantation. It is precisely the fact that an individual can be dead while many of his (her) organs can be alive and healthy that permits the transplantation of these organs into others. Most transplantable organs, including hearts, lungs, livers, and pancreases, become absolutely unusable if they are not removed before the donor's circulation stops. Although kidneys can recover from brief periods (up to about 15 minutes) of "warm ischemia"—remaining in the body at body temperature after cessation of circulation—the longer the period of warm ischemia, the longer the period before return of function of the transplanted kidney and the more complicated the care of the recipient. Thus, although it is possible to recover kidneys in the moments following circulatory arrest, there is a strong preference for obtaining the kidneys from a heartbeating donor.

The moral risk inherent in taking organs from a heartbeating, brain dead donor would not be worth assuming if there were not substantial moral benefits to be gained by the procedure. The societal benefits of organ transplantation are substantial and frequently underestimated by the public. For all organs the overall one-year graft survival now exceeds 60%, with the most common transplants having one-year graft survivals of 75–90%. For most organs, the five-year graft survival is in excess of 60%. Candidates for heart, liver, and lung transplants are expected to die within one year in the absence of transplantation, so a successful transplant in these cases is equivalent to prolongation of life. Rehabilitation following heart, liver, and lung transplantation is also excellent. The majority of successful recipients return to full-time work, homemaker status, or work as a full-time students. Since renal dialysis is alternative life-sustaining therapy for kidney failure, successful renal transplantation is not life-saving. There is convincing evidence, though, that transplantation leads to a better quality of life and better rehabilitation than dialysis. Treatment of renal failure with a transplant is substantially less expensive than with dialysis, and the calculated economic benefits of renal transplantation are substantial. The benefits of pancreas transplantation are more controversial. It is not yet clear whether this procedure affects patient survival or the development of the secondary complications of diabetes, but there is some evidence for better rehabilitation of diabetic patients in renal failure following combined pancreas/kidney transplantation than following kidney transplant alone, and many patients value highly the freedom from dietary restrictions and the need for insulin injections.

Since death by brain criteria is accepted by most of society, and since the benefits are real and substantial, retrieval of organs from brain dead, heartbeating cadavers is not very controversial where there is consent by the deceased and/or the next of kin. Similarly, the retrieval of tissues for transplantation is quite noncontroversial. There is no need to retrieve tissues such as corneas, bone, skin, vascular tissue, heart valves, and other connective tissues prior to the cessation of the heartbeat. These tissues are universally retrieved from heart-stopped donors unless they are obtained at the same time as an organ donation. Essentially, all religious groups, including orthodox Jewry, accept the legitimacy of this donation.

There remains considerable controversy, however, over the issue of who needs to give explicit consent for donation of cadaveric organs and tissues. This debate is fueled by the large gap between the number of individuals waiting for organs for transplantation and the number of organs that are currently retrieved from cadaveric donors. For the last several years, consent has been obtained for retrieval of organs and tissues from about 3,500 donors yearly. From these donors about 6,500 kidneys, 1,500 hearts and livers, and smaller numbers of pancreases and lungs have been medically acceptable for transplantation. Virtually all acceptable kidneys, hearts, and livers have been retrieved and transplanted. Whereas the number of organs retrieved has been fairly constant, the number of potential kidney and heart recipients has grown steadily so that now there are about 16,000 patients awaiting kidneys and about 2,000 awaiting hearts. It appears that the annual number of liver transplants has, similarly, now reached a peak limited by the number of donations. It is clear that the number of donors is inadequate to meet the number of potential recipients. It is estimated that the number of individuals dying each year in the US under circumstances in which they would be medically acceptable as organ donors is somewhere between three and six times the numbers that actually become donors. Whereas the reasons for the failure to identify and retrieve organs from a greater fraction of the potential donors are multiple, at least two issues relating to the consent process have been raised. First, occasionally, a family denies consent to donation even when the deceased has previously indicated willingness or signed a donor card. Second, given that most potential donors are the victims of sudden catastrophic events and are likely to be medically unstable, it is not uncommon that a potential donor dies before the next of kin can be identified and consent for donation obtained.

There are several schools of opinion concerning the minimum requirement for consent for donation. The most conservative position is that the next of kin, as the legal "owners" of the remains, must give consent. Contrary to this, the Uniform Anatomical Gift Acts in their original form (1987) make it clear that an individual can will his/her body, organs, or tissues irrespective of the preferences of the next of kin. These approaches both fall into the category of "opt in" rules; retrieval of organs and tissues is permissible only when permission is explicitly given either by the donor or the next of kin. At the other extreme, some European countries have recently adopted a presumed consent ("opt out") rule that says that organs or tissues can be obtained from a suitable donor unless the individual (or, alternatively, the next of kin) explicitly denies permission.

In practice, virtually all physicians in the US require permission of the family, despite the provisions of the Uniform Anatomical Gift Acts to the contrary. The reason for this conservatism of the US medical community is frequently misconstrued as fear of legal repercussions from a family if a decedent's organs or tissues are retrieved without their consent. This view of the situation misses the real human issues involved. As noted above, most of the time, the death of the potential donor is sudden and unexpected. The families are universally distraught, and most do not have a clear understanding of the nature of death by brain criteria and do not distinguish clearly between death and coma. It is hard for them to accept, in the pain of the moment, that their family member, lying warm on a bed with a heartbeat and (artificially maintained) respirations, is irretrievably dead. The medical personnel whose task it is to ask for consent are faced immediately with the distraught family, who may refuse to give consent, and not with the potential recipient(s) whose good lies with retrieval of the organs and tissues. It is humanly almost impossible to ignore the pain of the immediately present family to serve the needs of the recipients who are only abstract potentials. This is not to say that it is ethically justifiable to override the wishes of the deceased or the needs of society to appease those of the next of kin. It is simply to say that discussions of this issue that ignore the real human dimensions of the interaction are not likely to change the behavior of the health professionals.

A related question might be how often it occurs that a person is known to have given consent for donation, but this consent is overridden by the family, or how often opportunity to obtain organs or tissues is lost because of the delay to locate the next of kin and obtain consent. It is rarely known that someone has given consent to donation of organs or tissues unless this is mentioned by a member of the family. In most cases, the wishes of the decedent are honored by the next of kin. It seems unlikely, therefore, that more strict compliance with the terms of the Uniform Anatomical Gift Acts would have much impact on the number of donors. The situation with "opt out" laws is less clear. There are some indications from the countries that have passed these laws that the rate of donations is substantially higher than in the countries with "opt in" systems. This evidence, though, is still very provisional. At this point, it is not clear that a change in the ethical and legal requirements for consent would substantially change the rate of organ and tissue retrieval.

It was noted above that kidneys, unlike other organs, can survive brief periods of warm ischemia in the body of an individual who has recently died with a stopped heart. Rapid cooling of the kidneys to a temperature just above freezing protects them from the further effects of lack of circulation, and this cooling of the kidneys can be achieved *in situ* by injecting a cold solution into the aorta and into the peritoneal cavities via small catheters. The number of patients who die with a stopped heart in a hospital and who would otherwise be good kidney donors is considerably greater than the number who die with brain death and a beating heart.

The Anencephalic Infant as an Organ Donor

Anencephaly is an anomaly that occurs in the US in about three of every 10,000 births, or about 500–800 per year. In anencephaly, most of the top and back of the skull and scalp is absent, and there is no brain tissue above the brainstem. Specifically, there is neither any cerebral cortex nor any underlying cerebral white matter or nuclei. Given this defect, there is consensus that the anencephalic infant is permanently unconscious. On the other hand, it is not correct to say that the anencephalic infant is brain dead, or, as has has been suggested by some, brain absent. These infants have brainstems and their brainstem functions are intact to variable degrees. Thus, most anencephalic infants breathe on their own, suckle, and swallow, and many avoid painful stimuli, cry, and have facial expressions. These infants regulate body temperature poorly and have other abnormalities of brainstem function. They generally die from respiratory arrest within a matter of days after birth.

Interest in anencephalic infants as an organ donor has resulted from the combination of their permanent unconsciousness and abysmal prognosis with a serious lack of donor organs for transplantation into infants with surgically uncorrectable and life-threatening congenital heart disease. There is no clear need for retrieval of the lungs or pancreases of anencephalic infants, and most renal transplant centers prefer to use larger donor kidneys even for infant recipients. Although infant livers are also rarely available for transplantation, techniques exist for reducing the size of livers from a larger donor for use in a very small recipient. Conversely, there is no substitute for a size-matched heart donor, and, whereas the number of potential recipients for the hearts of anencephalic infants is not known, at least half the identified potential recipients die before receiving a heart transplant. Survival of infants following heart transplantation is as good as that of adults, so that a successful transplant leads to a real life saved. Therefore, the heart, and only the heart, of the anencephalic infant is a truly irreplaceable resource. Consideration has been given to waiting for anencephalic infants to die before using them as heart donors. Attempts to apply traditional criteria for brain death are made difficult by the absence of a skull and scalp and by the usually poorly developed central nervous system function of the newborn. Many of the medical criteria for diagnosis of brain death simply do not apply. Further, few, if any, of the organs retrieved from anencephalics after true brain death or cessation of the heartbeat have been successful. The above facts have all contributed to the argument that it is ethical to retrieve hearts from anencephalic infants prior to the occurrence of brain or cardiac death. These arguments have been supported by the pleas of several families who have sought to have their anencephalic infants used as donors to help make sense of what is otherwise a senseless tragedy.

However, it must be stressed that anencephalic infants are clearly not dead as defined by the Uniform Determination of Death Acts. At present, it would be illegal, and in many jurisdictions it would be first degree murder, to cause with premeditation the death of an anencephalic infant by taking out its heart. A change in this situation would require a change in the law specifically permitting this action. Several considerations have led members of the transplant community to recommend caution in moving in this direction. First, a law to change the definition of death to include the anencephalic infant, or to permit retrieval of organs prior to death of an anencephalic infant, would be motivated primarily by transplantation considerations, not, as in the case of the definition of brain death, for the benefit primarily of the donor and the donor's survivors. Many doubt the wisdom of authorizing such an action primarily for the benefit of someone other than the donor. Second, anencephaly is easily diagnosed early in pregnancy, and increasingly, families are opting for early abortion rather than carrying the pregnancy to term. Thus, the number of anencephalic infants will likely be smaller in the future. Further, there is concern that women, preferring an abortion, may be subject to subtle pressure to continue their pregnancies solely to give birth to an organ donor. Finally, there is concern that in the absence of complete public understanding and approval of organ retrieval from anencephalic children, the backlash against organ donation may result in a decreased number of donations from traditional donors, thus being counterproductive in net.

Use of Fetal Tissues for Transplantation

Probably the most hotly contested ethical issue in transplantation is whether it is permissible to use tissues obtained from abortuses for transplantation. This debate is tied to the debate on the ethical status of abortion itself. The basic question of those who oppose abortion is whether it is ethical to use resources obtained as a result of an unethical act even for highly salutary ends. For physicians, this is analogous to the question of whether the results of Nazi research on the effects of cold immersion, studies performed on prisoners and resulting in many deaths, should ever be published. There is no question that the information obtained might be useful, but publication might imply some degree of acceptance of what all agree to be immoral studies. The physician can contribute only a few considerations to this difficult debate.

1. There is, as yet, no well documented benefit to transplantation of any fetal tissue, though there are hopes that transplantation of fetal brain tissue may help patients with Parkinson's disease and that fetal pancreas transplants

may cure Type I diabetes. Therefore, the issue currently concerns research and potential benefits for humanity rather than known benefits.

2. Sufficient supplies could be obtained easily without any need to solicit abortions. It is unclear, therefore, that the use of fetal tissue would have any effect on the number of abortions performed.

3. To be useful, tissue must be obtained freshly and in a timed fashion. Unfortunately, tissue obtained from spontaneous abortions is not useful for these purposes.

4. It is quite possible that a cell line derived from a single abortus may become very valuable for transplantation to many recipients.

The question will have to be answered whether it is permissible to use tissue from a single abortus, perhaps from a medically indicated abortion, perhaps, for example, from an aborted anencephalic infant, when that use might benefit a large number of individuals.

Conclusion

A review of the medical facts underlying the ethical issues of transplantation may in some cases remove unfounded fears and make ethical decisions easier. A clear understanding of the medical facts of brain death, for instance, may provide reassurance that no harm is done to the brain dead donor. Unfortunately, it is as frequent that understanding the medical facts more clearly makes the ethical quandary more profound and more painful, as in the cases of anencephaly and use of fetal tissues obtained by abortion. We tend not to answer difficult ethical questions until we are forced by events to do so. The use of tissues from human donors for transplantation has raised many difficult ethical questions. Some of them have been answered or demand answers now; others do not have to be answered yet, but their time will likely come. In any case, though a clear articulation of the medical facts can provide the light needed to discern the answer, the answers must come from society as a whole, and not from the profession.

History of Transplantation and Future Trends

Todd L. Demmy
and George A. Magovern, Jr.

Early History of Transplantation

The story of transplantation is difficult to tell. Not only must one recount the work of many people striving toward the same goal, but one must also consider the multitude of other disciplines that transplantation embraces. These disciplines include antiseptic technique, renal dialysis, vascular surgery, organ preservation, immunology, cardiopulmonary bypass, intensive patient care, and the development of antibiotic therapy. Without the concomitant advancements made in these other areas, the field of transplantation would not be where it is today. Unfortunately, it is beyond the scope of this overview to detail these topics beyond their influence on the individuals responsible for the milestones in transplantation.

In the third century BC, two Chinese surgeons, Hua T'o and Pien Ch'isi, supposedly transplanted a variety of organs, but the historic accounts were quite vague and difficult to substantiate. In contrast, 400 years earlier in 700 BC, a clear technical description of a skin graft for nasal reconstruction was recorded in Sanskrit by ancient Hindu surgeons. The patron saints of medicine, Cosmos and Damien, are remembered predominantly for their fourth

From: *New Harvest* Ed.: C. D. Keyes
©1991 The Humana Press Inc., Clifton, NJ

century "miracle" of replacing the diseased leg of a parish-
ioner with a black leg from a Moor.

Otherwise, there are few historic records of transplantation work until the Renaissance. In 1547, Tagliocozzi began to reconstruct noses with skin still partially attached to
the forearm. He would sever the skin from its origin when
it obtained adequate growth and blood supply.

Modern history contains the majority of work in transplantation. John Hunter, the father of experimental surgery, performed his famous experiment in 1728. He
successfully implanted the claw of a rooster into its own
comb with good subsequent growth (autograft). Baronio,
in 1811, successfully implanted large autografts of skin
in other areas in the same animal. About the same time,
Brown-Sequard performed an interspecies transplant
(xenograft). Instead of a claw, the cock's comb now sported
a rat's tail.

Late in the nineteenth century, several transplant
methods arose that are clinically useful today. A German
surgeon named Thierock developed the split-thickness
skin graft in 1886. This allowed the donor site to heal and
be used repeatedly as is presently done for patients with
severe burns. The corneal graft was also developed and
was first performed in humans in 1905 by Zurin. To be
successful, the donor had to be a member of the same species (allograft).

The turn of the century also marked the turning point
of transplantation toward what it is today. Guthrie and
Carrel perfected many of the fine suture techniques in vascular surgery that are still in use. It then became possible to
move organs and maintain their blood supply. Between
1902 and 1905, multiple organs were transplanted between
animals by this team including thyroid, kidney, heart, and
even an entire dog's head. Clotting of the grafts were the
main difficulty and little mention was made about the phenomenon of rejection.

Organ Transplantation

Kidney

In 1902, prior to Guthrie's and Carrel's work, Emerich Ullman used magnesium tubes to connect the allograft kidney to the neck vessels of the dog with successful function. Unfortunately, the publicity of Carrel's work led to the transplantation of animal kidneys into human patients in the US (New York), Germany, and France between 1905 and 1910. All of these grafts failed and discredited the field of transplantation. Lexer, in 1911, demonstrated the poor survival of allograft kidneys (different animal of same species). Williamson in 1923 demonstrated the microscopic changes in rejected allografts. Humans were not involved again until 1933 when Voronary performed the first human allograft transplant. A cadaver's kidney was procured shortly after death and placed in the thigh of a Ukrainian patient with mercury poisoning. This and several subsequent patients' grafts were unsuccessful.

World War II led to the invention of the first dialysis machines by Kolff and Atwell, who donated them to various universities after the war. Merrill refined and developed the artificial kidney for use in Boston for patients with temporary renal failure and later for those patients with permanent failure who might undergo transplantation. In 1945, just before the machine became available, Hufnagel and Landsteiner transplanted a cadaver kidney into the arm of a woman comatose from acute renal failure. It sustained her for 48 hours, long enough for her own kidneys to improve, and she later made a full recovery.

In 1951, Hume, Thorn, and Dammen reported the first series of thigh kidney transplants to have any survival without immunosuppression. Then, 1953 marked the year of the first living related donor transplant. A mother donated a kidney to her son whose only native kidney was injured in an automobile accident. The allograft survived only 22 days. Finally, on December 23, 1954, Murray, in Boston, performed the first identical twin transplant. This was a landmark for several reasons. First, this was the first long-term success with restoration of health in the recipient. Second, the kidney was placed in the lower abdomen and urine drained into the bladder rather than through the skin, a technique similar to that used today. Third, it demonstrated that if immune factors could be controlled, transplantation could be successful. Finally, many important discoveries in immunology and the first successful use of the heart-lung machine occurred during this time period. Identical twins formed the majority of patients to undergo transplantation between 1954 and 1962. Murray (Boston) and Hamburger (Paris), however, treated a series of nonidentical recipients with total body irradiation as immunosuppression. This yielded few good results.

More modern immunosuppression was first used in April 1962, when Calne and Murray performed the first allograft transplant using the drug azothioprine (Immuran). The patient had an uncomplicated postoperative course that served to promote renal transplantation as a treatment with consistently good results.

Heart

The first animal heart transplant was done by Carrel and Guthrie in 1905 when a heart from a small dog was placed in the neck of a larger dog. Mann, a Mayo surgeon, continued this work in 1933 when he achieved an eight-day survival of a heterotopic (different than usual anatomic site) cardiac transplant. Mann also studied kidney rejection and stimulated work in liver transplantation. Experimental work persisted sporadically from 1933–1964. The Russian surgeon, Deminkov, performed the first heterotopic and orthotopic (normal site) cardiac transplants within the chest cavity.

It was not until 1964, three years before Barnard's widely publicized patient, that a heart transplant was performed in a human. A chimpanzee xenograft heart was placed in a 68-year-old patient dying from heart failure. The patient survived for only one hour after the operation because of the small size of the graft. It is interesting to note that a potential donor with irreversible brain damage was available and had been chosen for this patient. However, at that time there were no "brain death" laws. None of the transplant team could morally remove the donor from the ventilator so that heart function would cease; the recipient could not wait until the natural terminal event occurred.

Christian Barnard performed the first clinical heart transplant in August 1967, using a "brain dead" donor. His patient survived for 18 days until succumbing to pneumonia. Most of the experimental groundwork for this operation was done by Shumway and Lower at Stanford several years prior to this. They developed the operative technique that is used today and obtained good survival in canine heart transplants. The first clinical heart transplant after Barnard's was performed by this team.

By 1968, over twenty transplants had been performed in transplantation centers such as those in Paris and Houston. The poor results in these and subsequent patients led to the abandonment of cardiac transplantation, except by Shumway, who continued to study it. He documented the EKG and microscopic changes in cardiac rejection. Shumway eventually obtained a 50% one-year survival in about 40 patients by the mid-1970s. Barnard performed the first heterotopic cardiac transplantation in 1974 on a patient who could not tolerate a standard heart transplantation; however, this type of transplant never gained much popularity. Now, since the improvement in immunosuppression, an 85% first-year survival can be achieved and many centers are performing cardiac transplantation.

Liver

Welch performed the first experimental heterotopic liver transplantation in 1955. Earlier work was limited because of the complex blood supply of the liver. His technique did not require removal of the recipient's liver but the graft tended to atrophy if the majority of intestinal blood was not diverted toward it Also, the size of the "extra" liver in the abdomen tended to cause respiratory compromise.

Experimental orthotopic transplantation, therefore, was begun independently by Starzl and Moore in 1957. This led to the first clinical orthotopic liver transplant in 1963. The patient was a three-year-old child with congenitally malformed bile ducts (biliary atresia). Unfortunately, the patient died of hemorrhage on the day of surgery. From 1963 to 1967, there were continued failures in many transplantation centers including those in Boston and Paris.

The first success came in July, 1967, when Starzl performed surgery on an 18-month-old patient with liver cancer. The child died 13 months later from a metastatic tumor. By 1968, however, Starzl was able to report his first series of long-term survivors at the American Surgical Association meeting. The liver then became the first organ since the kidney to prolong life after transplantation.

Calne added cyclosporine A to the immunosuppression treatment in 1978 and obtained good results in his patients. Starzl added steroids in 1980 and achieved even better results. Now a 75% success rate can be expected with this therapy.

Immunosuppression

Immunosuppression is the practical application of discoveries in immunology to diminish the patient's resistance to foreign material. Its understanding progressed with each discovery in immunology and with each new transplant recipient. Only a few of the more important milestones can be discussed here.

Observations of rejection of foreign tissue have been made during the past several hundred years. Transplanted skin from another individual (allograft) would initially stick then slough in one to two weeks. Winston Churchill noted this in 1898 when he donated a small amount of skin to an injured fellow officer. When organs could be transplanted, it was noted that rejection occurred less often when the donor and recipient were genetically similar. Even then, rejection could occur with varied severity. Acquired immunity was not well appreciated until Holman (1924) clearly described the much faster rejection of a "second-set" of skin grafts after a first unsuccessful graft.

It was noted in 1945 that fraternal (nonidentical) twin cattle that shared the same circulatory system as fetuses were tolerant of each other's tissue. Medawar and Burnett extrapolated from this and made major discoveries pertaining to lymphocytes and clinical applications for which they won Nobel prizes. Tissue typing became possible by the work of Dausset, in Paris, and George Snell, an immune geneticist. This typing allowed a prediction of rejection to avoid poor transplant results. Terasaki brought tissue typing to the US and developed a test for preformed antibody to predict rejection.

Although total body irradiation was once used as a desperate means of immunosuppression, potent drugs are available today that are far more effective. The first significant experimental studies came in the early 1960s when 6-mercaptopurine was found to cause a delay in the disappearance of circulating antigen. Three men tested a derivative of 6-mercaptopurine: Calne perfected an animal model to test rejection; Hitching produced the drug with Burroughs-Wellcome; and Murray was the Peter Bent Brigham surgeon. This collaboration yielded azothioprine (Immuran), the first clinically useful immunosuppressive agent with which the landmark renal transplant patient in 1962 was treated. Starzl added steroids to Immuran to attain good results in 1963 but also noted the tumorogenic effect of immunosuppression. During this year, Goodwin showed that steroids reversed rejection.

Antithymocyte globulin was developed by Woodruff and its use reported by Starzl in 1967. In 1970, cyclosporine was developed by Borel and now a new drug is being tested that is 500 times more powerful with less side effects. Presently named FK-506, it may represent the next milestone in clinical transplantation .

Future Trends

It seems reasonable to predict that upcoming advances in transplantation will continue to be molded by forces affecting the field today. These forces include numerous educational, social, technological, and economic factors that decrease the number of organs available for transplantation. Examples of these factors include more stringent vehicle safety laws, better safety equipment (e.g., automobile safety bags), increased public awareness, and organ procurement expenses. The problem of fewer organ donors is compounded by an increased number of patients who are accepted as organ recipients by expanding or emerging transplant centers. We can assume that the dwindling donor/recipient ratio will necessitate maximal use of the solid donor organ and will lead to alternative means of organ replacement.

Because of the prevalence of heart disease, the number of potential recipients for this organ consistently exceed the supply, despite the high attrition rate for patients with severe cardiac ailments. Therefore, several centers are reexploring the possibility of using hearts from different species to augment or replace the failing hearts of human beings. Another interesting donor source has come into being with combined heart-lung transplantation. Since some of the recipients of these organs primarily have lung disorders, their excised hearts can be utilized for a different recipient. This "domino" transplant technology could possibly be applied to other organs as well.

The segmental anatomy of certain organs allow their division and use in more than one recipient. Initial successes in dividing both cadaver and living donor livers have been attained recently in Chicago and Pittsburgh. The use of heart valves as segmental organ transplants has shown good results as well.

The most promising future prospect is selective cellular transplantation. A number of techniques have been developed to harvest and culture only those cells responsible for the physiological activity that is deficient in the patient. If these techniques are ultimately successful, problems in organ availability may be alleviated or reduced because a large amount of tissue can be produced from a small sample of organ cells.

A great deal of experimental work has already been performed with diabetic patients and pancreatic islet cells. These cells, which are responsible for insulin production, comprise only a small portion of the gland. They have been separated from the remaining pancreatic tissue of one or more donors and injected into various areas of recipients (usually the liver vascular bed). Unfortunately, most of the results thus far have been unsatisfactory. Although it has been possible to reproduce cells in culture for some time, the problem has been to transfer these to the recipient in sufficient quantity to be effective. Most of the difficulty lies in providing an adequate blood supply to these clusters of cells. It may be possible to overcome this problem with bioabsorbable polymers like those found in resorbable suture material. It is possible to grow sheets of skin and liver cells on these polymer lattices, into which small blood vessels can rapidly grow after transplantation. Immature or fetal cells seem to be best for this process because of their high growth and duplication potential. Exciting laboratory experimentation is in progress that may lead to major medical breakthroughs. One example is the transplantation of bone marrow into fetuses while *in utero* in order to prevent later sickle cell disease. Another application involves the injection of immature heart cells into damaged cardiac tissue to effect repair.

The field of immunology is expanding so quickly that a discussion of possible future advances is beyond the scope of this essay, yet it is safe to surmise that a trend of ever greater selectivity similar to that of organ transplantation will prevail. Already, monoclonal antibodies are being used to induce graft tolerance. These antibodies have identical structure and kill specific lymphocytes that aid rejection. It is possible to apply selective treatment to the organ graft as well, thereby depleting cells that may attack the recipient. Transplantation of tissue with a great deal of immune cells, such as bone marrow or intestinal tissue, may improve with these methods.

For the near future, there are many types of organ transplants that are under investigation today but have not yet reached full clinical utility. Whole pancreas transplantation, popularized by Sutherland and Najarian, may prove useful to patients afflicted with diabetes that is difficult to control by insulin injections alone. Single lung transplants have been performed sporadically since the 1970s by Vieth and others for patients with emphysema.

We can expect many creative permutations in the number and types of organ transplantations. Multiple contiguous organ or "cluster" transplants are being studied for the treatment of complex abdominal diseases by Bahnson and Starzl in Pittsburgh. In addition to organs that are commonly transplanted, we can look forward to intestinal transplantations, as well as the utilization of other experimental organs. Small bowel transplant techniques were first described by Lillehei for patients with short bowel syndrome, but the large amount of immune tissue in the intestine makes this graft prone to rejection. The transplantation of connective tissues such as knee cartilage or teeth appears to show promise. Even the "final frontier" of the human brain has been breached. Donor nerve cells have been implanted into the brains of patients with Parkinson's disease.

There will probably be little visceral tissue remaining after organ procurement in future donors. However, alternative means of obtaining or producing organ tissue will probably have the greatest impact on large numbers of patients. It seems that any type of transplant is within the realm of human ingenuity. Hopefully, with continued work, more will become clinically useful.

PART THREE

Reproductive and Neurological Techniques

Reproduction and Transplantation

Ivan M. Naumov and C. Don Keyes

Our success has been greater in raising new,
than in answering old, questions.

M. C. Shelesnyak

New Reproductive Technologies

Reproduction can be viewed as a specific form or a "special case" of transplantation for the following reasons:

1. Transplantation can be defined as transfer of organ(s), tissue, or cell(s) from one biological individual to another;
2. The essential part that determines individuality, i.e., the biological identity of organ(s), tissue(s), and cell(s) is the genes;
3. The essence of reproduction is also the transfer of genes;
4. Therefore, reproduction can be viewed as a form of transplantation.

Of course, the major difference between reproduction and transplantation is that reproduction is a naturally occurring biological phenomenon whereas transplantation is not. With some of the so called "new reproductive technologies" (NRTs), however, artifice has been introduced into the natural process of reproduction. Although the degree of human intervention varies from one technique to another, all NRTs involve the noncoital or "artificial" procreation of life.

From: *New Harvest* Ed.: C. D. Keyes
©1991 The Humana Press Inc., Clifton, NJ

The introduction of NRTs into clinical medicine parallels the increased concern that society and couples have with infertility.* A widely accepted definition of infertility is the inability of a couple to conceive after 12 months of unprotected intercourse. This condition is present in 10–15% of the population of the Western world. Hull (1985) has reported that every sixth couple asks for a specialist's help to cure infertility. Within the last two to three decades, the increasing concern with infertility can be attributed to several causes. First, the post-World War II baby-boom generation was the first to be introduced to the potent new contraceptive, "the pill," and has employed it widely. This undoubtedly gave impetus to the liberation of sexual attitudes and to an increased number of cases of sexually transmitted diseases that, along with the increased number of abortions, resulted in a larger incidence of secondary infertility. Second, and of equal importance, is the fact that many women now postpone pregnancy, largely for professional reasons. It has been documented that fertility decreases with age. The third cause could be traced to the fact that one traditional remedy for infertility, adoption, is no longer as widely available as in the past. All of these, as well as the belief that technology can offer a solution to practically every problem, led to the pressure placed on the scientific and the medical community to seek ways to assist those who are unable to reproduce through sexual inter-

*This concern has increased during the last 20–30 years, although it had always existed in some form throughout the history of mankind. Fertility has been paramount to our race since its beginnings. Fertility in women had probably been worshipped even before the time of the first known symbol of fertility, the 27,000 year old figure of Venus of Willendorf. Fertility meant existence of the race. In Genesis, God' s instructions are to be fruitful and to multiply. Infertility has been considered a defect attributed almost without exception to the female. Talmud mentions three forms of living death: to have poor health, to be poor, and to be barren.

course. The following is a review of the reproductive technologies that are presently feasible, or are likely to be feasible in the future. They raise ethical questions that are not easily answered. Some of these questions will be addressed in the second part of this chapter.

Artificial Insemination (AI)

The artificial introduction of semen to the female reproductive tract for the purpose of achieving conception is the oldest NRT. This relatively simple technological process provides a first line solution to the problems of otherwise incurable male infertility. If the husband is the infertile partner, his wife can be impregnated by sperm from another donor (artificial insemination by donor, AID). This procedure can also be used if the husband is a carrier of a serious inherited disease or abnormality. If the husband's fertility is subnormal, e.g., if he has a reduced sperm count or if his sperms are relatively inactive, his sperms may be "improved" by laboratory methods such as "washing" to remove antigens that prevent conception. They can then be implanted in his wife artificially (AIH, where "H" stands for "husband").*

*AID and, to a much smaller extent, AIH have been used for many years throughout the world and today thousands of infertile couples have children who have been conceived through an AID program.

The practice of AID is opposed by mainstream religions. The Roman Catholic Church opposes the practice. The Church of England's position remains equivocal. It was instrumental in establishing one of the earliest committees to inquire into the practice in the United Kingdom. Its attitude has changed in recent years from condemnation to a lukewarm and reluctant acceptance.

The status of children born as a result of AID has been under discussion for some years. Since legislation concerning AID practice does not exist in many countries, the legitimacy of the child is determined by the courts on an ad hoc basis. The usual recommendation on this matter, including that given by The Council of Europe, is that a child conceived by AID with the consent of the mother's husband should be treated under the law, for all purposes, as the legitimate child of the husband.

In Vitro Fertilization

If both the husband and wife have normal sperm and ova but the wife's fallopian tubes are blocked or diseased, thus preventing passage of ova from ovaries to uterus, fertilization cannot occur naturally. There is yet another alternative that employs a bypass fertilization procedure. The first successful application of this procedure made history in July 1978 when a healthy female baby, later named Louise Brown, was born via cesarean section in Cambridge, England. Her conception was a result of a form of NRT known as in vitro fertilization (IVF), extracorporeal fertilization, or, simply test-tube baby procedure. Since 1978, IVF has become a landmark NRT and has generated enormous interest in society at large, as well as in the scientific and legal communities. Although newspaper photographs show the faces of happy couples and their IVF offsprings, societal debate concerning IVF continues unabated. Yet, despite significant opposition, this reproductive revolution continues, and is not likely to be stopped.

The technique used to conceive Louise Brown is now well known and is becoming a routine treatment for tubal and other forms of infertility. The infertile woman is given hormonal treatment to induce her ovaries to release more than one egg in the cycle. The hormonal levels and the

---

(continued from previous page)

In most countries where AID is practiced, the identity of the donor is unknown to both the prospective parents and their AID children. Even if they are informed about the circumstances of their conception, these children are not entitled to know the identity of their genetic father. Furthermore, The Council of Europe recommends that all precautions should be taken to keep secret the identity of all the parties involved, and even in court, the identity of the donor must never be revealed. Legislation in Sweden and Australia recommend that the child, upon reaching the age of 18, should have access to the hospital's records to obtain information concerning the ethnic origin and genetic health of his biological father. This was also the view of The British Committee.

development of the eggs are carefully monitored to detect the precise moment prior to the ripening of the egg(s) at ovulation. At this time, the egg(s) are retrieved by surgical methods and fertilized in vitro ("in glass," actually in a test-tube or petri dish) by the husband's sperm. If fertilization occurs, the embryo(s) are grown in an incubator for one to three days and then transferred via the vagina back into the uterus. The implantation and birth of healthy babies occurs in more than 20% of the cases in the most successful programs. So far, there has been no record of the feared increase in the rate of genetic abnormalities of IVF offspring. Also, there do not seem to be any negative biological consequences for the IVF patients as a result of clinical procedures. Therefore, most arguments against IVF remain ethical instead of purely clinical.*

Cryopreservation of Gametes and Embryos

In 1866, J. Mantegazza, a professor of pathology in Pavia, Italy, had proposed that banking of cryopreserved human semen might ensure the siring of offspring even after the father's death. The recognition of the practical value of his idea came much later. In the meantime, early in the twentieth century, commercial interests spurred the de-

---

*The major ethical concern about IVF arises from its unnaturalness with the result that, some argue, it should be rejected as a means of procreation. In a similar way, some claim that it is illicit because it involves masturbation. It also poses a risk for the offspring. Furthermore, some feminists claim that IVF allows increased male control over reproduction and hence threatens the status of women. As presently practiced, IVF does not disassociate marriage and reproduction because only couples are involved. However, since the capability of using sperm by a donor other than the husband obviously exists, ethical concerns about IVF include all of those found in AIH and AID. The objection to AID, namely that it dissociates marriage and reproduction by introducing a third party (the donor) does not really pertain to IVF, since in the majority of the cases, only the husband's sperm is used for fertilization.

velopment of procedures for the cryopreservation of the semen of domestic animals. Ultimately, human trials proved successful with the conception and birth in 1953 of a child conceived with frozen-thawed semen.

Sperm banking is now a very efficient routine component of every AI program. Cryopreservation of embryos and oocytes is, however, a relatively new procedure. Very few pregnancies have been established from frozen-thawed embryos and oocytes. It is to be expected that the success rates will improve. The storage of semen for future use (autopreservation) is primarily performed for men who are about to undergo a vasectomy or certain other medical treatments. Chemotherapy or radiation treatment of disorders such as testicular or renal cancer or leukemia may render a man sterile and are offered as reasons for autopreservation.

Oocyte and Embryo Donation

Oocyte and embryo donation (or embryo transfer, ET) are relatively recent developments in the rapidly expanding field of NRTs. These techniques have been feasible in animals for some time and the embryo transfer of high quality cattle embryos has, in fact, gained substantial commercial popularity as an animal breeding technique.

Oocyte donation is prescribed as a treatment to overcome female infertility due to ovarian failure or ovarian absence. Pregnancies brought to term after oocyte donation have been described by a few IVF/ET centers in the world. Unlike donation procedures for male gametes, the need for invasive techniques in oocyte collection for IVF/ET procedures presents a barrier to the rapid expansion of donor oocyte technology. The difficulties related to oocyte freezing and the limited number of oocytes theoretically and practically available limit the application and availability of oocyte donation. Normally, only one oocyte is released during a normal ovulation cycle and several are released during a stimulated one, but millions of spermatozoa are released during every ejaculation. The advent of ultrasonically guided oocyte recovery procedures (Lenz et al., 1982) and the new gamete and zygote intrafallopian transfer procedures (GIFT and ZIFT, respectively) introduced by Asch and coworkers in 1985 and 1987, should generally increase the number of excess oocytes available for donation. Pregnancies brought to term have been described using embryo donation following in vivo or in vitro fertilization (Lutjen et al., 1984; and Bustillo et al., 1984). The development of oocyte and embryo donation procedures are based on improvements in cryopreservation techniques. Successful cryopreservation of human embryos helps overcome the waste of spare embryos resulting from hyperstimulation therapies in IVF/ET programs. This development might eventually overcome the ethical objections of those who believe that every fertilized ovum should be treated as a human being with a right to life.

Surrogate Motherhood

Surrogacy is the practice whereby one woman carries a child for another woman with the intention that the child should be handed over after birth. The use of AI or IVF have eliminated the necessity for sexual intercourse in order to establish a surrogate pregnancy.

The surrogate arrangement is normally discussed as a transaction between the surrogate mother and an infertile married couple that wants a child. The medical indications for surrogacy based on a woman's infertility are rather limited. In some instances, the woman is not capable of carrying the baby for various medical reasons. Instead of adopting a genetically unrelated baby, she may want a baby that is genetically half her husband's.

Surrogacy raises different ethical questions due to its very nature. First, it separates childbearing and child rearing functions and, thus, dissociates physiological motherhood from social or legal motherhood. Second, it evokes the fear that women will be exploited and children turned into commodities. Is surrogacy a form of prostitution or a morally responsible act of controlled reproduction? Is it the sale of a baby or of reproductive services? Does the money passing from hand-to-hand offend our moral senses? What about the basic rights of women to use their bodies in their own ways?*

Male Pregnancy

In the mid-sixties, the fertilized egg of a female baboon was transplanted into the abdominal cavity of a male baboon. The embryo attached itself to the omentum and formed a placenta, the organ that would normally attach to the uterus and draw nutrition from the mother's circulation during pregnancy. The embryo grew in the abdominal cavity of the male baboon for four months before it was surgically removed. The delivered fetus was reported to be alive, although the pregnancy was not carried to term (seven months).

To date, this baboon transplantation is the only experimental proof of the feasibility of male pregnancy. The idea has been around for some time, but no one has tried to implant an embryo into a man's abdominal cavity where the embryo/fetus would have nourishment, grow to term, and be delivered by an operation similar to a cesarean section.**

*The legal problems of surrogate motherhood stem from the fact that the mother giving birth is the legal parent of the child. The surrogate's legal parenthood must be terminated and then the child has to be adopted by the contracting parents. The surrogacy agreement between the parties has no legal effect, although, in 1987, a New Jersey court decided differently in the widely publicized case of Baby M. The genetic mother was given some maternal rights.

In the UK, the Warnock Committee recommended criminalization of surrogate motherhood agreements. In Denmark, most surrogate arrangements, whether for profit or not, are forbidden as of 1986.

**What about the candidates for prospective male "mothers"? According to an Australian poll, the candidates would be primarily homosexuals,

Numerous technical and social difficulties would accompany a human male pregnancy. The first, and most obvious, one is the availability of spare embryos. It is very unlikely that there could be a large pool of donatable embryos for such a procedure. The priority demand for spare embryos would certainly come from infertile couples. Even if there would be embryos available for male implantation, the failure risk of the surgical embryo transfer procedure would be high. Hormonal support would probably be needed, both for implantation and for providing a hormonal status to the male "mother" comparable to that of a pregnant woman. If the fetus successfully survived the nine months in the abdominal cavity, the delivery could be another dangerous stage of the pregnancy. In women, the uterus, which is a strong muscular organ, contracts after the delivery and shuts off the blood vessels eroded by the placenta. Since there would be no uterus in the male's abdomen and the abdominal wall would not contract, this could lead to a massive and fatal hemorrhage after the surgical delivery.

Reproductive Organ Transplants

There is no reliable evidence of successful transplantation of female or male reproductive organs in humans. However, Andrews (1984) mentions that a man in the UK, born without a testicle, had achieved a pregnancy with his

---

(continued from previous page)

heterosexual men who had infertile wives, and single men who wanted to fulfill their need for a child. Dr. John Money of Johns Hopkins University originally envisioned only one kind of person that would be an applicant for a male pregnancy: the male-to-female transsexual. For this population, "it is so terribly important to experience everything a woman can experience." Can this psychological need also be found in other men, homosexual or heterosexual, who want to bear children? Why would heterosexual men, married or single, want to become parents this way? We simply do not know. We do know that male homosexual couples often want children. Male impregnation could be appealing to them.

wife after he received a transplant of one testicle donated by his brother. Andrews also reports that De Cherney and coworkers have attempted to transplant fallopian tubes from one woman to another, but the operations haven't been successful. An attempted ovary transplantation has been stopped by British health authorities who claimed that under existing laws any resulting offspring would be considered illegitimate. Even if ovary transplants were allowed, their value would probably be limited. A postmenopausal woman, for instance, would not be able to conceive after receiving an ovary transplant from a younger woman. Research in rats has shown that ovaries, transferred from young fertile rats to old infertile rats, take on the pattern of the old rat. The ovary from an old rat, however, can be activated by transfer to a young rat. This may mean that a woman who was of reproductive age but who lacked ovaries or whose ovaries were defective, could receive an ovary transplant from an older woman (one who is, for example, undergoing a hysterectomy) and, thus, have a chance at achieving a pregnancy.

Research on Human Embryos and Fetuses

Studies on early human embryos and fetuses are likely to improve current methods of clinical treatment. Edwards (1989) claims that research on human embryos should only be devoted to studies that promise considerable clinical advances:

1. Studying the origin of human chromosomal disorders;
2. Growing human embryos in vitro and ectogenesis;
3. Typing embryos; and
4. Using embryonic cells and tissues for transplantation.

The fourth area is of special interest to the general topic of this book.

Indeed, one of the most exciting and promising applications of embryo and fetal tissue research has been the use of fetal cells. In a recent case, fetal cells were used to treat another fetus *in utero*. The French physicians, Jean-Louis Touraine and Daniel Raudrant, treated a 30-week-old fetus, diagnosed with a rare and frequently fatal immune deficiency (bare lymphocyte syndrome), by injecting immune cells from the thymus and liver of two aborted fetuses into the umbilical cord of the deficient fetus. Second trimester fetal liver contains a rich source of hematopoietic stem cells, universal precursors of all the blood cells. At this stage, the donor fetus immune system has not yet developed and normally histoincompatible stem cells may be transplanted into the immunodeficient fetus to produce reconstituted immune cells. The postnatal tests have shown that some of the grafted cells have seeded and multiplied in the infant's spleen, liver, and bone marrow. The infant is alive with a partially functional immune system. Equally astonishing are some of the results and prospects of fetal cell transplantation as a therapeutic tool for neurotherapeutic disorders such as Parkinson's disease, as the next chapter explains.

In principle, it should be possible to establish cultures of fetal cell-lines from different organs that can be grown in vitro for a long period of time and serve as sources of transplantable precursors of vital organs or tissues. One of the first such cases will probably be that of bone marrow transplant, since there are established methods for its long-term culturing.

Biotechnological Realities and Perspectives

Cloning

Cloning is the production of biologically identical individuals. All the cells of these individuals possess exactly the same set of genes and the organisms, as a whole, are expected to be morphologically and functionally identical. Cloning may be achieved in two different ways. The first method is embryo fission, in which an oocyte is fertilized in vitro and then divided (at one of the 2- to 32-cell stages) into two or more new embryos and finally transferred to the uterus. Each newly formed embryo would be genetically identical to the others. In case of a successful pregnancy, this would lead to multiple births of identical monozygotic twins, triplets, and so forth. This procedure has been successfully performed in frogs, sheep, and cattle. The second method is "true cloning." This involves replacing the genetic material (the male and the female pronuclei) of one freshly fertilized egg with a diploid set of chromosomes from any bodily cell, then initiating cleavage and the later formation of a viable embryo, fetus, and offspring. So far, this method of cloning has not resulted in the birth of cloned animals, although early formation of embryos has been accomplished.

Gene Transfer into Human Gametes and Embryos

For more than a decade, reproductive biologists have introduced foreign genes into animal embryos. In many instances the foreign genes, introduced chiefly by microinjections or by retroviral carriers, have been successfully expressed in the offspring, and then carried on to the next generation. In this way, "transgenic animals" have been created, possessing biological traits not inherited from their parents. The tools of this type of genetic manipulation have so far been used not only to generate precise animal models for human genetic diseases, but also to produce agriculturally useful animals, as in the case of the transgenic pig, implanted with a growth gene from a cow, resulting in a higher meat-to-fat ratio than conventional pigs. The prospects of introducing normal genes to embryos, or even to gametes prior to fertilization, in order to replace defective or substitute missing genes, are quite realistic. If applied to humans, this form of germ-line gene therapy could become a powerful clinical weapon for the ultimate cure of presently incurable inherited diseases. Cystic fibrosis, Huntington's and Tay-Sachs diseases, sickle cell anemia, muscular dystrophy, and many other diseases in the list of 4,000 known genetic diseases could thus be prevented. Surely this would be beneficial. However, these techniques might also provide the means of generating children with superior physical and mental traits. It is one thing to use the techniques to remove defects. It is another to use them to produce "better" human beings. Inevitably, many questions will be raised: What is a better human being? Is it a better athlete or a person with a higher I. Q.? Might not a Down Syndrome individual be a better person than a diabolical genius? Who will determine these criteria?

Gene Transfer as a Model
of the Crisis of Unnaturalness

Gene transfer is an extreme model for most of the ethical issues found in the other NRTs, and it reveals a peculiarly critical relation between consequentialist and deontological ways of approaching respect for life. Consequentialist concerns typically include both the immediate and long-range effects as well as the possible risks or benefits of the procedure. Gene transfer involves two distinct areas of deontological concern: (1) the status of the embryo fertilized in vitro, and (2) the obvious unnaturalness of altering the genetic structure, hence self-identity, of human beings. This second concern in gene transfer pushes the crisis of unnaturalness to its limit.

Status of the Embryo

Like other NRTs that include IVF, gene transfer raises deontological concerns about the duty to protect life: When does human life begin? Most of the fertilized oocytes fail to survive and decisions must be made about what to do with those that do survive but are not implanted. The three interpretations of the status of the fertilized oocyte in "Beginning and End of Biological Life" (this volume) are that it:

1. Has human rights since it is essentially human at conception;
2. Is an object without rights until birth; or
3. Becomes human at some point between conception and birth.

Ethical concerns about the beginning of human life are critical only for those who hold the first position and for some who hold the third.

Those who hold the first position will object to the high failure rate, since fewer IVF embryos survive compared with those fertilized naturally. This objection might be mitigated by the fact that the natural reproduction process destroys perhaps 70% of the embryos conceived within the body. If this is correct, nature's 30% survival rate only slightly exceeds the 15–25% success rates sometimes claimed for IVF. At the same time, gene transfer appears to add risk not found in more conventional uses of IVF, namely, an extremely high failure rate. "True cloning," as defined above, adds yet another and possibly ambiguous dimension to this concern, since it would destroy the displaced genetic core of the host embryo into which the new genetic material from a donor embryo is transplanted. The displaced genetic material is approximately at that stage in which the male and female pronuclei are merging. Concern about the destruction of this material is obviously most acute for those who believe that the embryo is human at conception.

Those who hold the third position, that the embryo becomes human at some point between conception and birth, differ from one another as to whether the life of the embryo is worthy of respect prior to the decisive moment at which it becomes actually human. This volume's "Beginning and End of Biological Life" considers the argument that the embryo becomes fully human with the beginning of brain function, but that it is worthy of a type of respect at conception because that is the point at which it becomes "unconditionally potentially human." Accepting this or any argument that attributes "partial humanness" to embryos between conception and brain function makes research on embryos and the risk factor of gene transfer more ethically problematic and less susceptible of resolution than accepting the first or second positions.

Crisis of Unnaturalness

All NRTs, without exception, are unnatural. Sexual unnaturalness, however, appears to cause more concern than other types. This is undoubtedly due to a particular historic attitude toward sex, which holds that sexual relations can only be justified by the conception of offspring. If conception is the natural end and only good purpose of sex, any interference with this process is wrong. Some assume that there is an uncompromisable (deontological) duty to respect and protect natural processes because they are sometimes taken to be essential to the order of things as created by God. Gene transfer involves two distinct types of unnatural intervention: dissociation and modification of genetic self-identity.

Unnatural Dissociation

Gene transfer dissociates certain structures that belong together in what some consider the right order of nature:

1. Intercourse and ejaculation;
2. Reproduction and intercourse; and
3. Marriage and reproduction.

With respect to the third point, gene transfer dissociates marriage and reproduction only if the donor of a gamete for IVF is other than a spouse. While current practice limits IVF almost exclusively to married couples, nonmarital donors are theoretically possible in gene transfer as well as other uses of IVF. Such a dissociation of marriage and reproduction would not be unnatural in the biological sense. Some will consider it unnatural, however, in another sense, if they accept marriage as the only appropriate authorized means of reproduction. The first and second points refer to biological unnaturalness.

Since gene transfer requires the donation of sexual cells, it potentially dissociates intercourse and ejaculation. Whether it actually does depends on how the sperm is obtained. Some consider masturbation unnatural, hence unethical, even if done for the sake of procreation. This concern could be eliminated by marital intercourse with a condom to catch the sperm for AIH. Some might consider even this use of a condom unnatural (owing to its association with contraception). It has been suggested that a way to avoid this particular objection might be to use a condom with a hole in it, small enough to trap enough sperm for subsequent insemination but large enough to provide the fiction of naturalness. Gene transfer necessarily dissociates intercourse from reproduction as do all procedures based on IVF in the sense that conception takes place outside the body and is initiated by a member of the IVF clinical team, not the parents.

Unnatural Intervention into Genetic Identity

Gene transfer is more unnatural than IVF in the sense that it intervenes at the subcellular level and artificially changes the genetic self-identity of the individual and possibly the species. Speculative procedures like cloning, which might be possible in the medical future, seem to exemplify this to a high degree. Gene transfer for the purpose of gene therapy might be technologically possible in the near future. Are there limits that should be placed on "artificiality" as such? If so, is therapeutic gene transfer within those limits?

Respect for life, the foundational principle of biomedical ethics, seems at least initially to conflict with itself in the case of gene transfer both deontologically and consequentially.

A Deontological Thesis. Artificial intervention into the human genetic structure for any reason, including therapeutic, is unethical because it interferes with natural reproduction processes. Gene transfer into embryos violates the genetic identity of the individual and species. The same essential objection applies to gene transfer into gametes prior to IVF. Reproductive processes should follow the right order of nature. Respect for life consists partly in preserving that sanctity of natural reproductive processes.

A Deontological Antithesis. Eliminating disease and improving the quality of life through artificial intervention into natural reproductive processes is ethical because it respects the sanctity of life. This respect consists partly in producing the kind of genetic combinations that will prevent suffering both in the individual and in the species. The deontological principle of beneficence and avoidance of harm is deliberately unnatural when it tries to restore health or prevent disease, in the sense that disease is a natural phenomenon. The fact that an event is natural is no guarantee that it is good, and reproductive processes that perpetuate disease are no exception. Apart from the question of the embryo's status in risky procedures, the deontological antithesis in favor of therapeutic gene transfer is consistent with the consequentialist concern about whether it achieves its intended effects.

A Consequentialist Thesis. Artificial intervention into the human genetic structure is ethical if it has beneficial effects. This consequentialist stipulation that intervention must fulfill its purpose agrees with the deontological antithesis favoring artificial genetic intervention to prevent disease. The consequentialist argument is based on the assumption that the procedure is likely to benefit a human life that has intrinsic value, and on the belief that it is proper to use human powers to do this.

A Consequentialist Antithesis. Arguments against the possible adverse effects of gene transfer arise from at least three sources. First, it could be argued that the present level of scientific knowledge does not assure that gene transfer can be done without undesirable genetic side effects. Second, allocating resources to develop gene transfer might not be equitable in view of other, more immediately pressing needs in preventing and healing disease. Furthermore, gene transfer might be used to benefit rich, otherwise privileged, groups to the exclusion of the poor and minorities. Third, there is the danger that genetic engineering will not be limited to correcting genetic defects, but that it will be expanded in undesirable ways. To take the most radical case, gene transfer might not be limited to a given species, but human and animal hybrids might be produced through interspecies transfer. It is even more likely that gene transfer could be used to improve the human race, but the standards of improvement might be arbitrary, based on fallible judgment or unconscious bias or even on the private interests of those who control the procedure.

Conclusion

Ethical reflection should avoid both uncritical permissiveness and thoughtless opposition to biotechnology as applied to human reproduction and the treatment of genetic diseases. A more rational approach is to focus attention on the type of regulations that ought to control the practical application of these new techniques. In fact, one can offer an ethically positive justification for such intervention to prevent disease, a justification that is both deontological and consequentialist. Genetic intervention "to improve" the human species, however, is a completely different issue. Although there are possibly no purely deontological objections to attempting it, there are serious problems of a consequentialist type. The most obvious is the danger of abuses like those mentioned above. It is rational to set certain limits to the use of our powers and to restrict gene transfer to procedures that are not ethically ambiguous. If we begin by recognizing that genetic intervention is probably inevitable as a supplement to natural mutation in shaping the future of human life, then attention can be focused on properly restricting its use. The human species seems to have evolved to the point that human control will partly determine the future of evolution.

Brain Tissue Grafting

James E. Wilberger, Jr. and C. Don Keyes

Transplantation into the Brain

In contrast to intrinsic nerve fibers of the brain and spinal cord, peripheral nerve fibers can regenerate and form functional reconnections after injury. Neurobiologists have been attempting to elucidate the differences in regenerative potential between central and peripheral neurons since the end of the last century. As a result of the extensive research into central nervous systems (CNS) regeneration in the 1960s and 1970s, it was definitively proven that central neurons have the capacity to grow, elaborate processes, and establish new synaptic connections long after the embryonic growth period is over. This capacity, known as collateral sprouting or reactive reinervation, was shown to be a general property of most, if not all, intrinsic central neurons. However, collateral sprouting occurs only in intact neurons in response to nearby injury; damaged central neurons do not regenerate. Although collateral sprouting may allow for the reestablishment of synaptic connections between neurons, these connections are not necessarily either anatomically or functionally correct.

Based on the information concerning the ability of the CNS to regenerate, questions began arising as to whether replacing injured central neurons with neurons of like kind would allow for the reestablishment of anatomical and functional integrity in damaged neuronal circuits. Thus, the concept of the intracerebral transplantation of central

From: *New Harvest* Ed.: C. D. Keyes
©1991 The Humana Press Inc., Clifton, NJ

nervous tissue evolved. Within the past five years, this has become a widely used means of studying the ability of central neurons to regenerate and form functional reconnections after injury. Although the majority of CNS transplantation research remains in the domain of the neurobiologist, a surprising amount has already reached the level of practical clinical applications. The capability of transplanting neurons to correct CNS abnormalities, whether due to disease, trauma, or genetic defects, has enormous implications.

Historical Perspective

The ability of various tissues to survive and grow after transplantation into the CNS has been known for a long time. In the 1890s, Thompson claimed to have successfully transplanted pieces of mature cat cerebral cortex into adult dogs. The transplanted tissue appeared viable on histological section seven weeks after implantation. In the 1940s, Le Gros Clark published convincing results about the survival of transplanted embryonic rabbit cerebral cortex. The tissue became organized and established connections within the host rabbit cortex.

On a somewhat grander scale, in the 1960s, White et al. showed that it was feasible to isolate and remove the entire brain and its circulation without seriously compromising function—a necessary first step for organ transplantation of any type. White and his colleagues isolated 63 monkey brains and developed the extracorporeal systems necessary to maintain them in a viable state. They then used the information thus obtained for the actual transplantation of a whole dog brain preparation into the neck of a recipient dog. Grafted whole brains were viable for periods ranging from six hours to two days. The persistence of significant electrical activity of the brain was used as a measure of viability. Failures of the transplants were due to low blood pressure in the recipients or to compromise of blood flow from the transplants. Histological section failed to reveal any cellular abnormalities that might be compatible with rejection phenomena.

In the late 1960s, Deminkov, a Russian investigator, studied the transplantation of a whole head together with the surrounding tissues and organs. Twenty such experiments were carried out; 19 of the transplanted heads survived for periods of one to 29 days. Deminkov indicated that in all cases after the recipient dog awoke from anesthesia, the transplanted head also awoke.

The transplanted head reacted briskly to the surroundings, had an intelligent expression, licked when it saw a bowl of milk, and eagerly lapped up milk or water. When the recipient dog stood up, and the transplanted head experienced discomfort or pain, it bit the recipient dog on the ear. The transplanted head fell asleep irrespective of whether the recipient dog was awake or asleep.

Interest in CNS transplantation was revived in the late 1970s. The exceptional reports by White et al. and Demikov notwithstanding, the available literature to that time implied that, whereas the survival of neuronal transplants was possible, the ability of the transplanted tissue in the host brain to establish structurally accurate and functional interconnections was highly questionable. Nevertheless, intensive investigation by many researchers around the world addressed and proved several hypotheses that formed the basis for the current clinical application of CNS transplantation. These researchers definitively showed that

1. Transplanted neurons can survive in a host brain, develop processes, elongate, and form synaptic connections;
2. Transplanted neurons can organize into anatomically correct configurations and establish functionally correct synapses;
3. Synaptic connections can become electrophysiologically or secretorially functional.

Once this information became available, various researchers began considering the practical clinical application of CNS transplantation in various disorders.

Transplantation for Parkinson's Disease

A syndrome described by James Parkinson in 1817 is among the most common and readily recognized of neurological entities and is responsible for great disability, particularly among the elderly. Although Parkinson syndrome is associated with a variety of causes and pathologic conditions, a great majority of cases have a relatively uniform clinical and pathological appearance and no evident cause and are termed idiopathic Parkinson's disease. Patients with Parkinson's disease suffer a gradual loss of motor function with akinesia, rigidity, trembling, gait disturbance, and loss of postural reflexes. The associated pathological findings include a loss of nerve cells from pigmented brainstem nuclei, particularly the substania nigra and the locus ceruleus.

Little real progress was made in the first 140 years after the description of Parkinson's disease either in understanding its pathophysiology or in developing effective treatment. A radical change occurred in the late 1950s with the discovery that there are high concentrations of the catecholamine neurotransmitter dopamine in normal basal ganglia. In brains of Parkinson patients, a correlation was made with cell loss in the substantia nigra and depletion of dopamine from the basal ganglia leading to the concept of this disease as a dopamine deficiency syndrome. Since dopamine does not pass the blood brain barrier, the early treatment trials, based on that concept, focused on the use of the dopamine precursor Levodopa. Levodopa and the more recently introduced receptor agonists are effective in the treatment of Parkinson's disease, particularly in its early course, and represent substantial improvement over previous therapy. Nevertheless, it is now clear that Levodopa and receptor agonists are only limited ameliorative treatments and that the course of most patients is one of inexorable progressive disability regardless of therapy. Furthermore, the treatment is usually associated with a series of unpleasant and distressing side effects, including drug induced dyskinesias, on/off phenomenon, and psychiatric disturbances. For a number of years, despite the undoubted benefits of current treatment, Parkinson patients and their doctors have been frustrated and unhappy. It is within this context that brain transplantation in animal models was begun. An animal model of Parkinsonism was developed in the 1970s by destroying neurons of the nigrostriatal system with a neurotoxin-6/hydroxydopamine. This produces a syndrome of akinesia and rigidity, the hallmarks of the human disease, in animals and, strikingly, can be reserved by transplantation of fetal substantia nigra to the neostriatum. The transplanted tissue was shown to grow and reinervate the neostriatum and functional restoration could occur despite the fact that the transplant was not in a normal location and did not have normal connections with the remainder of the brain. Researchers in Sweden subsequently discovered that the transplantation of adrenal medullary cells would also produce a functional recovery in a Parkinson animal that was similar, if not identical, to that produced by the transplantation of fetal substantia nigra. These observations led to an attempt by Swedish groups in 1894 to transplant adrenal medullary tissue into the brain in two patients with Parkinson's disease. This study was carried out in patients with far advanced disease. The adrenal medullary tissue was stereotatically implanted deep into the neostriatum. The transplanted tissue, however, had little, if any, clinical effect in the patients treated. In 1987, researchers and clinicians from the Hospital de Especialidades Centro Medico "Laraza," Mexico City, reported dramatic improvement in two Parkinson's disease patients as documented by detailed neurological evaluation, as well as demonstrated on video tapes, led to an almost immediate reaction in the US and the establishment of multiple programs for transplantation for Parkinson's disease. Since that time, more than 200 patients have received brain transplants with adrenal medullary tissue for Parkinson's disease in 15 centers around the US. However, despite the information that has been accumulated, it is still uncertain whether this procedure has lasting benefits for patients severely affected with Parkinson's disease. None of the investigators in the US have yet been able to duplicate the dramatic improvement reported by the Mexican researchers. Nevertheless, some beneficial effects have been seen in patients receiving the adrenal transplants. It appears that a certain number of patients seem to respond better to antiparkinsonism drugs that may not have been working previously. In a few, the amount of time that they are incapacitated is reduced and the quality of their life seems to be improved overall. Nevertheless, no one has been cured and all still require medication for their disease.

Embryonic Tissue for Brain Transplantation

Investigators at this time are beginning to question whether adrenal tissue may be the right choice for transplantation in Parkinson's disease. As noted previously, early laboratory investigations indicate that the embryonic brain tissue from the substantia nigra may be the more appropriate choice because it may eventually differentiate to produce dopamine neurons. Additionally, in other types of disorders such as stroke or trauma, if brain function can ever be restored by transplantation it is most likely going to be necessary to replace the damaged neurons with similar, functional neurons. Embryonic brain tissue would appear to be the most likely candidate since this is the only type of brain tissue that still has the capacity to grow, differentiate, and develop functional synaptic connections.

The two main obstacles to the clinical utilization of embryonic brain tissue are donor selection and the phenomenon of tissue rejection. Embryonic tissue would have to be obtained from either a fetus or a neonate. Stillborn fetuses might provide a source of tissue; however, stillbirth generally results from major congenital abnormalities, infections, or lack of blood flow or oxygen to the fetus. Thus, when stillbirth occurs, the brain is dead and there are no viable tissues to harvest for transplantation. The only options thus available would be to consider the use of brain tissue made available from induced abortions. In 1986, a conference was held at Case Western Reserve University on the ethics, law, and science of neural transplantation. It was the consensus opinion that human embryonic cell grafting could be performed legally in the majority of states utilizing existing guidelines for cadaver tissue for organ donor programs. By 1990, though, the medical community has yet to come to grips with the use of fetal tissues for this purpose.

The use of neonatal brain tissue raises a separate set of concerns. Once birth occurs, the declaration of death requires strict adherence to a rigid set of criteria (*see* "Beginning and End of Biological Life," this volume). The declaration of brain death requires the complete absence of neocortical and brainstem function. Thus, it remains controversial whether brain tissue obtained under such circumstances would prove viable for transplantation purposes.

In spite of such concerns, the clinical utilization of fetal brain tissues for transplantation is proceeding. In early 1988, it was reported at the University of Lund, Sweden, that human fetal embryonic tissue from the substantia nigra was implanted into the brains of two patients with Parkinson's disease. The researchers reported no improvement in these patients. Fetal grafting has also been undertaken in London, although no results have been forthcoming. Similarly in the US in 1989, two institutions undertook fetal intracerebral grafting for Parkinson's Disease with mixed results. As a result of growing interest in this area of research, the National Institutes of Health held a consensus conference on the use of human fetal tissue from induced abortions for research purposes. Taking into consideration the ethical, legal, and scientific questions involved, the conference petitioned the secretary of Health and Human Services (HHS) to allow the advancement of such research. However, after six months of further review, HHS recently issued a moratorium on further fetal research in light of growing national concerns over abortion-related issues.

A possible alternate source of transplantable neural tissue would be in the development of in vitro cell lines by pooling cells from several donors or by genetic cloning. Work has already been done with a type of brain tumor cell, neuroblastoma, which also produces multiple neurotransmitters. In the laboratory it is possible to render the tumor cells amitotic—unable to proliferate—yet still have the cells produce neurotransmitters. A major concern over the use of genetically-engineered cell lines or inactive tumor cells would be the possibility of the cells growing unchecked inside the recipient brain and having a destructive rather than beneficial effect.

Once acceptable tissue for brain transplantation has been established, another obstacle is that of transplant immunology. When organs are transplanted, the body's immunological system views the transplant as foreign, potentially harmful material and attempts to destroy or reject it. For kidney, heart, and other organ transplants, tremendous strides have been made in the use of potent immunosuppressive drugs that can block this phenomenon. The brain has been traditionally viewed as an immunologically privileged site. This view holds that the brain is incapable of mounting an immunological reaction against foreign materials, thus making it a safe haven for transplants. Nevertheless, the immunology of neural transplants has not been carefully studied. The few autopsy studies from the Swedish and Mexican adrenal-implant studies indicate that the autologous adrenal tissue has degenerated and been replaced by scar tissue. Whether this was due to immunological rejection is unknown. The few patients who have received fetal transplants are being treated with immunosuppressive drugs.

Conclusion

The tremendous scientific and public interest in neural transplantation research at the present time gives optimism that the obstacles of tissue availability and immunology will be overcome, leading to the possibility of a broader application of neural transplants for neurologic disease and injury. Currently, however, there is tremendous and appropriate skepticism in the scientific community based on the to-date experience with adrenal transplants for Parkinson's disease. Many are calling for more controls on the current clinical application of CNS transplantation with respect to subject selection criteria, benefits and risks, uniformity of surgical techniques, duration of follow-up, and ethical considerations. Nevertheless, it appears possible that, for specific neurologic disorders, CNS transplantation may evolve into an effective, readily available, and acceptable treatment option.

Ethics of Transplantation into the Brain

CNS transplantation travels the wavering line between experiment and innovative treatment. It is subject to the ethical limitations placed on both by medical ethics. Still, it has promise. In addition to Parkinson's disease, animal research suggests potential human application in at least three other areas, as Wilberger (1983) suggests.

First, research into peripheral nerve-axonal bridge autografts raises the hope of partially repairing human brain (and possibly spinal cord) damage. There is some evidence that these grafts might induce regrowth of injured neurons themselves, not just collateral sprouting. If regrowth occurs, then there is the further and unsettled question of whether it is capable of establishing functional connections. Despite these uncertainties, there are grounds for speculation about partially repairing some types of brain damage. Wilberger, in the article just cited, says that research "has demonstrated that it is possible to stimulate mature, supposedly deviant intrinsic central neurons to regenerate in a manner that closely resembles regulating peripheral nerve fiber."

Second, it might eventually be possible to correct neuroendocrine deficiencies by transplantation into the brain. Fine (1986) writes that

> In contrast to neurotransmitters, the hormones that are secreted by certain structures in the brain normally travel considerable distances to their targets, and their release may be regulated not by signals from adjacent cells but by diffusible factors carried in the blood or released by other brain structures. Neuroendocrine diseases,....are therefore promising candidates for treatment with neural grafts. The first such disease in which the technique has been tried in the laboratory is diabetes insipidus...

Third, research into the transplantation of hippocampal tissue raises the possibility of one day developing a treatment for Alzheimer's disease. According to Fine,

> Experimental models of Alzheimer's disease offer another striking example of transplant-mediated recovery. The progressive loss of memory and other higher mental functions...is associated with widespread degeneration of neurons and neurochemical abnormalities, notably a depletion of acetylcholine in the hippocampus and much of the cortex.

Fine reports that researchers have

> ...transplanted solid pieces of tissue from the forebrain structures that normally supply the hippocampus with acetylcholine into the cut fimbria of rats; in other rats the workers introduced the tissue directly into the denervated hippocampus...The grafts improved the animal's ability to learn a maze...

CNS Transplantation to Restore Function Does Not Transplant Consciousness

To clear up a common misunderstanding, it must be said that none of the types of CNS transplants mentioned above involves the transplantation of consciousness (self-identity). Personal traits are not transferred from donor to recipient. However, CNS transplantation to restore function is sometimes confused with speculation about (1) the transplantation of whole heads from one animal body to another and (2) the transplantation of whole brains or parts of brains, extracranially. The first is theoretically possible within the horizons of today's technology, and is adequately and properly criticized by Wiest (*see* "Christian Perspective," this volume). The second is presently not even hypothetically thinkable within the present limits of our rudimentary knowledge of the brain, but speculation about it nevertheless persists. Miller (1971) writes

> Should the brain become a transplantable organ, the donor would become the recipient. Should Mr. X die of a condition which apparently had not affected his brain, and this brain were to be transplanted into the vegetable-like body of Mr. Y, then Mr. Y, upon rising to the commands of this new brain, would be Mr. Y in physical appearance only, for his brain, the source of life, thought, movement, and memory, would be Mr. X. Thus, in being the donor of a brain, Mr. X would actually become the recipient of the body of Mr. Y. Hence, in actuality, this would be a body transplant, rather than that of a brain. Thus, Mr. Y, with all of his previous physical characteristics would be in reality Mr. X. Upon arising from his hospital bed, he would return to the home of Mr. X; he would respond to the name of Mr. X. He would pass by the wife and children of Mr. Y and physically embrace Mrs. X and call the X children as his own.

The very thought of grafting any amount of adult brain material to transplant consciousness pushes the crisis of self-identity to its limit, even though all this is merely speculative. A part of that crisis is its tendency to block our attempts to formulate the ethical issues that surround it, even though Wiest seems to identify certain important points. Speculation about brain transplants and tissue grafting to transplant consciousness force us to look the brain-mind problem in the face even when we succeed in denying it elsewhere.

Ethics of CNS Transplantation*

At least two basic ethical issues arise from the treatment of Parkinson's disease and the three potential types of CNS transplantation to restore function mentioned above: the use of embryonic and fetal tissue and respect for the recipient's life.

Embryonic and Fetal Tissue Sources

The use of embryonic and fetal tissues is the most controversial issue related to CNS transplantation at this time. Drawing information from Vawter et al. (1990), we should distinguish three distinctly different types of issue, ranging from least to most controversial, according to the source of human fetal tissue.

SPONTANEOUS ABORTIONS AND SO ON. The use of tissues from spontaneous abortions, ectopic pregnancies, and still-births are ethically acceptable to most persons. However, as Vawter et al. report, "these sources are generally in very small supply, are available only sporadically and unpredictably, are of dubious quality and safety, and are often less than scientifically optimal for research and therapy."

ELECTIVE ABORTIONS FOR OTHER PURPOSES. The use of tissues from dead fetuses after the fact in elective abortions is controversial because it might encourage abortion or lead to other abuses. Mahowald et al. (1987) refer to typical

*Ethical issues surrounding neural grafting will be discussed in a report being prepared by the Congressional Office of Technology Assessment, which will be released in 1990 or early 1991.

"slippery slope" arguments of those who have concerns of this type:

> As with many troublesome ethical issues, the slippery slope argument is applicable to transplantation of fetal tissue. For example, we may initially permit only the transplantation of tissue from dead fetuses. If this does not prove successful or adequate, we may then transplant tissue from nonviable (living) fetuses or abortuses. Routinization of the practice could lead to transplantation of larger and larger portions of the brain, to transplantation of entire brains from viable fetuses, or to harvesting organs from other donors who are not dead, but are dying or chronically ill.

There is the further danger that using fetal tissues from elective abortions will lead to commercialism. Mahowald et al. also mention that:

> ...while the technology might at first be used purely for convincing therapeutic reasons (to heal or save lives in situations of seriously debilitating disease, for example), morally questionable motives such as profit making could eventually take over. That possibility has of course existed throughout the history of organ and tissue transplantation. Here one could envision a situation in which women would be paid to become pregnant and undergo abortion exclusively for the sake of fetal tissue transplantation. The moral problems thus raised parallel those that may occur in surrogate motherhood.

Despite serious consequentialist concerns like these, the use of tissues of dead fetuses from elective abortions does not seem to conflict with the "at conception" and "unconditionally potential" positions about when human life begins as long as the abortion is not performed for that purpose. For example, Kelly (*see* "Christian Perspectives," this volume) mentions the argument of those who claim that "once the abortion has taken place and the aborted fetus is dead, the organs and tissue may be taken as with any other dead person." Concerning the danger of commercialization,

Mahowald et al. point out that reliable checks are possible to prevent this kind of abuse. Such wedges "are neither new nor ineffective; they have been successfully applied to organ and tissue retrieval from cadaver donor." This underscores the need for public policy that will establish checks against abuse.

LIVING FETUSES. Procuring tissue from living fetuses is a separate issue altogether and includes at least three considerations.

First, one must determine whether the fetus is alive or dead. Vawter et al. report that

> ...it is currently not possible to establish brain death in fetuses. We must rely on the heart-lung criterion for the determination of fetal death. This criterion is sufficient and can be applied without ambiguity in the vast majority of cases of fetal death due to spontaneous or elective abortion. However, if abortion techniques are modified so that aborted fetuses are not dead according to the heart-lung criterion when tissue is procured, or if tissue procurement techniques are modified so that tissue is procured from living fetuses prior to abortion, a moral justification for tissue procurement from living nonviable fetuses becomes necessary.

Second, *in utero* and *ex utero* techniques are distinguished. Both have been used to procure tissue from fetuses presumed to be living, as evidence cited by Vawter et al. suggests:

> There are reports of modifications of abortion techniques that remove the fetus from the uterus intact keeping the fetus from being disrupted and making it easier to procure certain types of fetal tissue...A report from Sweden describes a technique that involves removing tissue from a fetus *in utero* immediately prior to a suction abortion. Such changes in abortion and tissue procurement procedures may increase the chances that tissue will be procured from living fetuses.

Third, a distinction is made between viable and nonviable fetuses. Fetal viability, according to Vawter et al. is "an empirical indicator that the probability of fetal survival outside the womb is greater than zero." This depends primarily upon the degree to which the lungs are developed. According to law, a viable fetus *ex utero* has "the same right to treatment as any term newborn" and it may not be killed or allowed to die in order to procure its tissues. This report cites two approaches to procurement from nonviable fetuses, namely (a) reclassifying them "as dead or not yet living," thereby changing the definitions of death or life, or (b) allowing tissue removal while considering them living, thereby changing "the presumption that a human being must be dead before tissues can be removed."

"Reclassifying" nonviable fetuses according to approach (a) would depend upon the definition of brain death. According to the "higher" definition, "a non-viable fetus *ex utero* that exhibits heart beat and respiration, yet has irreversibly lost the potential for consciousness, is dead." According to the whole-brain definition, "a nonviable fetus *ex utero*, that has no evidence of brain-generated integrated functioning and has permanently lost its potential for such functioning is not alive." The question of nonviable fetuses *in utero* becomes complex to the extent that "these fetuses retain the potential for consciousness or for whole-brain life."

Allowing tissue to be removed from nonviable fetuses, while considering them alive according to approach (b), is advocated by some. Vawter et al. explain that those who hold this view do so without becoming involved in the reclassification problems mentioned above:

> They see no compelling ethical worries relative to the nonviable fetus that cannot be managed relative to other moribund and uncomprehending living donors who have no interests, who have permanently lost all interest, or whose interests are outweighed by the interests of others.

The ethics of tissue procurement from living fetuses depends on when human life is believed to begin. Those who hold it begins with conception will consider all such procurement wrong. If human life begins at some point between conception and viability, such as brain function, then procurement from living fetuses is wrong after that point. Some brain life theorists will consider it permissible prior to brain function, whereas others would claim it violates the rights that an unconditionally potential human being ought to have. The distinction between *in utero* and *ex utero* seems to have no deontological significance (except for possible danger to the mother) as soon as the embryo or fetus reaches the point at which it is considered human. Of course, in all of the distinctions just mentioned, the line is drawn prior to viability. A viable fetus has a *prima facie* claim on life and the means to sustain it. There is good reason for the legal focus upon viability, even though its timing will vary with advances in biomedical technology. Still, there are ethical considerations that argue for drawing a line at an earlier stage of fetal development.

Respect for the Recipient's Life

CNS transplantation to restore function must respect the life of the recipient. The deontological concern for life and the consequentialist concern for good effects on human life come together. This convergence of deontology and consequentialism sets limits to risky procedures in medicine.

Respect for life includes the integrity of self-identity. Whereas the transplantation of brain tissues to restore function does not involve the substitution of one self for another, any alteration of the body can have some effect on the self. Alterations in brain functions due to implantation of tissues might have consequences not yet anticipated. Even more, the brain has come to symbolize our consciousness of self. This might affect recipients of CNS transplants. It also probably affects what the general public thinks or feels, and this, in turn, could go far toward determining public policy.

Symbolism of the Brain

Brains differ from other vital organs not only anatomically, but also in terms of what they represent. Faces surround them and eyes are external extensions of the brain. At the same time, the face and the eyes are organs by which one is recognized and identified. They are instruments of personal identity and its expression. Head injuries that damage the brain often disfigure the face. Surgical access to the brain usually means opening the head. The organ of self-identity ironically happens to be housed behind the face, which is the most obvious visible symbol of self-identity. Gorman (1969) claims that the peculiarity of the image of the brain comes from two paradoxes. The first is the fact that our image of the organ of perception is not based on perception:

> While the hand's appendages, the fingers, enable us to feel the hand, and the eye may see itself, one's own brain has not been touched, nor has it been felt, even by the most curious... Not only are we denied the possibility of touching our own brains, but also the brain itself is imperceptive to touch...Yet, the brain functions as a center for discrimination by means of touch as well as by many other modalities of perception.

The second paradox comes from the way in which the "imperceptive brain" is encased "by a highly perceptive head":

> ...vision constantly informs us of the presence of the head, and touch and other senses give us many stimuli that arise from the head. But immediately inside our perceptive heads lie our imperceptive brains...The contrast between the perceptivity of the abdominal wall and its contents is enormous...which instantly brings to awareness a reminder of the perceptive void within.

Unlike our images of other parts of the body, which are "formed by the interaction of perception with conception," our inability to perceive the brain forces us to base our conceptions of it on the perceptions of specialists. Such conceptions of the brain, Gorman claims, are then "richly overlain with our fancies and or fears." Brain symbolism seems to be in a state of conflict with itself, and concern about transplantation into the brain might be at least on the periphery of this conflict.

PART FOUR

Psychiatric and Philosophical Perspectives
on Self-Identity

Psychological Perspectives on the Process of Organ Transplantation

Edward E. Kern

Introduction

Organ transplantation is not a single event. To the patient, transplantation surgery is only one of many events on a continuum reaching back to the first sign of illness and stretching forward into an uncertain future.

Transplantation has taken on an aura of uniqueness that obscures the fact that, first and foremost, it is a medical treatment with the same goals as any other medical treatment: the preservation of life, the restoration of health, and the relief of suffering. Dramatically improved energy, mobility, and emotional well being are the rewards for many patients who receive a transplanted organ. Life is not only extended, but a process of restoration occurs in which a healthy "new" organ replaces an "old" diseased one. Less fortunate patients find themselves in limbo as the risks and complications of transplantation prove all too true. These individuals must strive for a meaningful and stable life despite continued ill health. For a few, the process is simply a prolonged state of dying.

Each patient must cope with such unavoidable stressors as the known risk of dying before receiving an organ,

From: *New Harvest* Ed.: C. D. Keyes
©1991 The Humana Press Inc., Clifton, NJ

the demands of a strict medical regimen, and the possibility of organ rejection. These universal challenges give rise to corresponding psychological reactions. For example, most patients feel anxious as they face the uncertainty of waiting for a donor organ, and they feel transient mood depression following the first diagnosis of organ rejection. The psychological outcomes of transplantation, however, range across the spectrum of human experience. The personality and social milieu of each patient powerfully shapes that individual's response to circumstances. Viewed from this perspective, there are as many outcomes as there are patients.

For this reason, then, no one psychological model can fully describe the complex tapestry of emotions, thoughts, and behaviors observed in a patient. Each model does provide an essential framework for organizing observations and asking significant questions. In addition, a valid model should permit some predictions about the most likely range of a patient's emotions, thoughts, and behaviors under given circumstances. For example, the clinician using a diagnostic model asks whether a patient's depressed mood is a symptom of a psychiatric illness called "depression," or simply a normal reaction to events. An affirmative answer—diagnosis of the illness—predicts a response to medical treatment for depression. A negative answer suggests that ordinary emotional support may be all that is needed for the mood to lift.

This chapter will examine the human dimension of transplantation from the viewpoint of five psychological models: diagnostic, psychodynamic, coping-with-illness, quality-of-life, and family system. Each model provides a distinct vantage point for looking at a patient's passage through the transplantation process. Taken together, like pieces of a puzzle, a picture of the patient's psychological and social functioning can emerge.

Five Psychological Models

The Diagnostic Model

A system of psychiatric diagnosis classifies abnormal thoughts, emotions, and behaviors into discrete categories separate from a continuum with normality. This follows the traditional medical model. A persistently sad facial expression, for example, is viewed as a sign of an illness called "depression" rather than an extreme example of a normal emotional state.

The diagnostic model holds great value for the clinician whose goal is to identify and treat psychiatric illness. Numerous studies demonstrate the power of certain diagnoses, such as "Major Depression" or "Panic Disorder" (*Diagnostic and Statistical Manual of Mental Disorders*, Third Edition, Revised, 1987) to predict a response to specific treatments.

Organ transplantation was still in its infancy when the first observations were published describing the occurrence of serious psychiatric disorders during the course of kidney and heart transplantation. Episodes of agitation, psychosis, or severe depression occasionally preceded or followed transplantation surgery. The psychiatrist could readily classify these phenomena using the accepted diagnostic model. Fortunately, these severe disorders were often responsive to clinical treatments—medication and psychotherapy—identical to those applied outside the transplantation setting.

One of the most important functions of a diagnostic system is to provide a common language for clinicians to communicate about patients. Confusion occurs when the name of a diagnostic category is the same as the name for a common emotional state. For example, the word "depression" can mean either a common emotion known to everyone or a specific treatable psychiatric disorder. The person who observes a sad and tearful patient facing a second transplant operation may naturally say, "I'd be depressed, too!" without stopping to consider that he or she is using the common, not the diagnostic meaning of the word depression.

The clinical literature suffers from this same confusion. Few studies have rigorously applied standard diagnostic criteria to clearly discriminate psychiatric illness in transplant patients from variants of normal emotional states. Thus, the reader cannot always decide whether a published report describing the emotional distress seen in many transplant patients should be understood to reflect an abnormal occurrence or simply the expected course of events. In fairness to clinical researchers, the distinction can be quite difficult. Typical symptoms of psychiatric illness, such as fatigue, insomnia, tremulousness, vague bodily discomfort, or loss of appetite, are also common in many other medical conditions. Since psychiatric diagnosis is currently based more on classifying signs and symptoms rather than causes, a high degree of diagnostic certainty is not always possible.

What is known is that patients who undergo transplantation are at least as vulnerable as the general population to serious psychiatric disorders. Debilitating depression, suicide, psychosis, and drug or alcohol dependency can occur following the most successful transplantation. The challenge to the clinician is to identify preoperatively the patient at risk for postoperative psychiatric morbidity. Prevention or early intervention are then possible.

The patient who has already suffered from a psychiatric disorder such as depression or drug dependency is at higher risk for the same disorder than would otherwise be the case. This assumption is based on the natural history of such disorders in the general population. Whether or not transplantation actually alters the long-term prognosis for these disorders is unknown. Conversely, the occurrence of even serious psychiatric illness does not necessarily preclude a successful transplantation.

The Psychodynamic Model

The psychodynamic model evolved from Sigmund Freud's original theory that mental phenomena are shaped by the dynamic tension between instinctual "drives" and internal or external constraints on these drives. The small child's obvious desires for special attention, omnipotence, and protection persist as adult fantasies just outside conscious awareness. In the normal adult personality such infantile desires, barred from direct satisfaction, are channeled into socially acceptable forms by more mature processes such as empathy, self-discipline, or humor. To a point, these mature qualities characterize each individual's responses to the demands of daily life.

The strain of life-threatening illness and transplantation causes each patient to regress toward more childlike ways of thinking and acting. For example, the desire to be shielded from harm by an all-powerful parent may re-emerge as a tendency to elevate the physician to a god-like status as protector—a temporary source of comfort to a patient. Conversely, the patient deprived as a child of the opportunity to experience consistent trust may be provoked to fear and anger by the experience of placing his or her life in the hands of another, despite rationally knowing the necessity of doing so.

Psychodynamic theory is also consistent with other personal experiences reported by transplantation patients at one time or another. For example, many patients report intense emotional turmoil as the childlike illusion of personal immortality confronts, despite silent prayers and promises, the reality that their own organs must be replaced to extend life. In addition, many patients describe feeling guilty at the thought that "someone must die for me to live." This reveals a persistent infantile fantasy that merely thinking about something may cause it to happen: transplantation as murder.

The psychodynamic model posits the existence of psychological "defenses," such as denial, projection, or humor, to account for the many ways individuals respond to anxiety-provoking thoughts and events. Emotions are viewed on a dynamic continuum with each patient's vulnerabilities and defenses intimately involved in shaping the moment-to-moment mood state. Thus, one may speak of anxiety "breaking through" the defense of denial in a man awaiting heart transplantation who suddenly shakes with fear after his roommate is wheeled away to receive his transplant first.

The psychodynamic model has merit when the clinician needs to understand the individual patient. In contrast to the diagnostic model, the psychodynamic model avoids classifying patients into descriptive categories. Its rich and detailed vision of psychological processes comes closer to answering the question "Who is this person?" than it does to answering the question "Does this person have a diagnosable psychiatric illness?"

A number of early reports on the psychology of kidney and heart transplantation revealed intriguing fantasies that, for the first time, allowed the clinician to look at the way in which the sense of self and body image depends not only on external appearance, but on the perception of internal body organization. Some organ recipients believed they would somehow take on personal characteristics of the donor. These "incorporation fantasies," studied from a psychodynamic perspective, revealed that a new organ is not a psychologically neutral addition, but rather must be integrated into a revised sense of personal identity. For example, Castelnuovo-Tedesco (1981) reported the following interesting case:

> A man received a kidney from a donor known only to be a 38 year old male. The recipient imagined the donor was married with children and therefore must be a stable person. The recipient then concluded that he had received wisdom and common sense from the (cadaver) donor.

It would be natural to wonder if all recipients believe they will take on attributes of the donor. Only a small minority, however, have persistent or maladaptive beliefs of this type. Incorporation fantasies are most accurately viewed as interesting but psychologically benign.

The Coping Model

A large body of clinical literature examines the myriad ways people attempt to cope with the demands of being sick. For example, Lazarus and Folkman (1984) define coping as the "constantly changing cognitive and behavioral efforts to manage external and/or internal demands that are appraised as taxing or exceeding the resources of the person." These authors divide coping efforts into two types: "problem focused" attempts to master external demands, and "emotion focused" attempts to maintain emotional stability. This coping model can be applied to better understand the psychological and behavioral aspects of transplantation.

Each patient is endowed with certain psychological resources. For example, intelligence, flexibility, a sense of humor, and the ability to push unwanted thoughts out of mind serve the transplant patient well as he or she confronts such challenges as waiting for a donor organ, complying with the strict medical regimen, resuming normal sexual activity, or grieving the death of a fellow transplant recipient. An example of "problem-focused coping" is seen in the patient who actively seeks information on the best type of exercise program in anticipation of returning to work after a heart transplant. "Emotion focused" coping is exemplified by the pretransplant patient who talks with the hospital chaplain after witnessing the sudden death of a roommate who was also on the waiting list for an organ.

The degree of perceived stress, and, therefore, the demand on coping resources is specific to the individual. One patient might judge postoperative pain to be quite tolerable, yet feel overwhelmed at the prospect of taking immunosuppressant medication faithfully for the rest of his or her life. Another person may find taking medication to be little problem, but at the same time have difficulty grieving the loss of attention that accompanies a return to health after a long period in the sick role.

A patient's efforts to cope may lead to unintended consequences. For example, as a way of reducing anxiety a patient might blindly place his or her trust in the clinical staff to the point of not wanting to know much about the upcoming transplantation surgery. Whereas this posture might minimize anticipatory anxiety, the patient will be at a distinct disadvantage due to lack of information when he or she recovers from anesthesia in a half-awake and confused state. Unrealistic expectations might then be associated with bewilderment and difficulty coping with unavoidable postoperative pain. Continued indefinitely, this effort to control anxiety through the use of denial becomes ever more harmful as the pattern of avoiding full comprehension of information continues into the post-hospitalization phase.

The coping perspective dovetails with the commonly accepted paradigm of psychological "stages"—pre-operative, postoperative, and post-hospital—through which every transplantation patient must pass. Much of the clinical literature has been organized around this stage concept, which reflects as much about the sequence of demands of the transplantation process as it does about the intrinsic nature of the patients.

The emotional challenges of each stage are predictable and fall into two basic categories: uncertainty and loss. Each patient faces the uncertainty of waiting for a donor organ that may never come. He or she must simultaneously prepare for life and death. Optimism that an organ will eventually come must not obscure the necessity for putting one's affairs in order. The prayer for rebirth competes with the uncertain vision of saying "good-bye" to loved ones. After successful surgery and the resumption of life activities, the specter of rejection is a constant companion. Loss of energy, job satisfaction, sexual attractiveness, personal relationships, financial security, or dreams of growing old with one's spouse are only a few of the losses facing many patients at one stage or another of the transplantation process.

Coping is specific, not global. A transplant patient may report that he or she is coping well in the sense of having a high quality-of-life, but in reality the individual responds to challenges one at a time. The focus of coping efforts may shift from minute-to-minute. For example, the patient engaged in a "problem focused" effort to plan financial strategy may quickly shift to an emotion focused effort to minimize anxiety as he or she is suddenly distracted by worry that normal sexual functioning will never return. Gentle humor about the fragility of life may lighten the burden of hard-nosed decisions about exercise, diet, and stress management as the patient copes day-to-day.

The Family System Model

The family system model assumes that illness encompasses not only the patient but all significant persons in his or her life. From this viewpoint, an isolated examination of the individual is meaningless. Emotions, thoughts, and behavior are to be understood in the context of family interaction and social feedback.

A useful example of a family system model for conceptualizing the human dimension of illness is described by Rolland (1988). Rolland postulates that the impact of an illness on the patient and his or her family is determined more by the onset, course, outcome, and degree of incapacitation than by the specific medical diagnosis.

For example, the onset of an illness may be slow and the course progressive as with chronic renal failure. Or, it may be abrupt and episodic as with repeated myocardial infarctions. The outcome of an illness ranges from the restoration of full health to death. Incapacitation ranges from minimal to total.

This model is based on the assumption that every family progresses through a life-cycle with each successive stage bringing new challenges for each family member. In general the tasks of one phase are expected to be moving toward completion before the next phase is encountered. Life threatening illness like any unexpected upheaval, disrupts the family life-cycle.

A fictitious example will clarify the application of this model:

> An eighteen year old boy, the youngest of three children, has recently gone off to college. His parents, having completed the child rearing phase of the family life cycle, now look forward to more time alone to renew intimacy within their marriage. One month later he develops a viral cardio-myopathy, which rapidly progresses to the point at which a heart transplant is required. This previously vigorous young man, who is just entering a developmental phase of increased independence, suffers intense emotional distress associated with regression to an almost child-like state of dependency. His parents are immediately thrown back into the role of taking care of a dependent child, a role that they had not needed to fulfill for many years. An older sister, whose wedding is only months away, feels intense guilt as she anticipates moving away and beginning her own family at a time in which she clearly sees her family of origin plunged into a crisis. The father, who had been planning to reduce his work hours, now fears financial ruin and begins seeking additional overtime.

This case illustrates severe psychosocial disruption of the entire family system caused by an incapacitating illness with an acute onset and possibly fatal outcome, which struck at a vulnerable point in the family life-cycle. The addition of another stressor, such as job layoff, legal problem, or alcohol abuse could trigger a downward spiral seriously threatening the family's integrity.

Had the illness struck at a less vulnerable point in the family life-cycle, or had it been less sudden and incapacitating, the family might have adapted with less disruption.

Suppose the same young man is twenty-eight rather than eighteen years old, and has a family of his own. He and his spouse directly confront the crisis of his sudden illness. His family-of-origin, however, is less directly involved than would have been the case had the transplant been necessary when he was eighteen years old. His parents will undoubtedly provide emotional support, yet they will be less likely to regress into a caretaking role. The patient himself is now in a later phase of his own life cycle, defined by much greater independence from the family of origin. His parents will likely react to his illness more as a challenge than a duty. Each family member still experiences emotional distress, but severe disruption of the family life cycle is less likely.

In contrast, imagine the illness had been one that was less incapacitating and of slower onset, such as chronic renal failure. Now, the parents are not unexpectedly plunged into the caretaker role. The sister's plans are not disrupted. The patient himself remains in college. Thus, despite the need for an eventual kidney transplant, he can still leave behind adolescent dependency and identity issues and move into the young adult role. Each family member remains "on track" in the life cycle and there is less chance of any person becoming overwhelmed with emotional distress.

The advantage of the family system perspective lies in its ability to depict a transplant patient's psychological state in terms of the many external forces that influence it. This model helps explain certain changes in emotion, thought, and behavior that cannot be accounted for by studying only the individual.

The Quality-of-Life Model

The term "quality-of-life" (QOL) appears with increasing frequency in the clinical literature on transplantation. As a theoretical construct, the QOL model assumes that measures of emotional well being, daily activity, vocational rehabilitation, and social functioning can be combined into a global assessment of a person's life.

Among the various perspectives on the psychosocial dimensions of transplantation the QOL model has employed a more quantitative approach. For example, the concept "Quality-Adjusted-Life-Year" (QALY) uses an estimate of the quality-of-life to adjust the measured length of survival after transplantation (O'Brien et al., 1987). A shorter period of survival may be counter-balanced by a higher quality-of-life during the survival period to give a score (QALY) that is equal to that of a longer period of survival associated with a lower quality-of-life.

The quality-of-life perspective serves goals other than the clinical care of the individual patient. The QOL model shows its power as an aid to decision-making, as in the allocation of health care resources, for example. This model can also serve as a tool of communication between clinician and patient when a decision must be made between alternative treatment approaches to a potentially fatal disease.

Judgments about the outcome of medical treatment have traditionally been based on the unspoken assumption that a longer life is necessarily of greater value than a shorter life: the goal of medical care is, in fact, to postpone death. The very idea of a "Quality-Adjusted-Life-Year" challenges this assumption. Use of this model is equivalent to accepting the premise that a trade-off between quality and quantity of life is not only useful, but ethically acceptable.

Obviously, the QOL perspective cannot hope to capture the rich psychological detail that is revealed to the clinician who explores from a psychodynamic viewpoint the life of the individual patient after transplantation. Nor can it help the clinician decide whether a particular patient's depressed mood should be understood as a manifestation of psychiatric disorder or simply as a normal phenomenon.

But it may aid the policy-maker grappling with the decision of how many dollars should be shifted away from other health care needs to cover transplantation services.

From a QOL viewpoint, the psychological outcome of transplantation is not constrained by diagnostic categories. A 40-year-old man who achieves three years of extended life to enjoy his family and part-time job might be judged a more satisfactory outcome than a man of comparable age who lived twice as long following transplantation, but whose postoperative life was characterized by unemployment, severe marital discord, and social withdrawal. These two very different outcomes could occur without either patient ever experiencing a clinically diagnosable psychiatric illness.

Studies using questionnaires to assess QOL consistently document that a large majority of persons who survive transplantation report being highly satisfied with life. Vocational rehabilitation has generally been reported to be good, regardless of the type of transplantation, with most recipients working or in school.

An important but unanswered question is the degree to which reports of life satisfaction are biased. Although patients are aware that the answers to questionnaires are anonymous, there is ample evidence from other psychological studies that the "social desirability" (Crowne and Marlowe, 1960) of an answer does influence the likelihood that an individual will objectively reveal his or her true feelings. A very high life satisfaction rating may be an unconscious attempt to say "thank you" to the transplant team.

The fact that so many recipients, regardless of the type of transplant, rate their satisfaction as high also raises the possibility that the very act of giving the answer may serve to emotionally bolster the individual's decision to have undergone the transplantation in the first place. This hypothesis is consistent with the act of "bolstering" observed in an extensive study of decision-making in all walks of life

(Janis and Mann, 1977). Janis and Mann demonstrated that after arriving at a decision, most people no longer give equal weight to evidence contrary to their choice. Thus, having made the decision to extend life by accepting the gift of an organ from another person's body, the recipient would naturally be biased away from conscious awareness of a negative outcome.

Summary

Choosing a Model

The five theoretical models just discussed represent a sampling of the ways one can think about the psychology of the patient undergoing transplantation. Additional models are available, and several are mentioned in the next section.

Before selecting a particular model, one must ask, "What exactly do I want to know?" and "At what level of complexity do I need to know it?" Only then is it possible to choose an appropriate model as a basis for further inquiry.

The psychiatrist or psychologist working with individual patients frequently uses both the diagnostic and psychodynamic models. These models aid the clinician in making complex treatment decisions for individual patients.

The social worker, an essential member of every transplant team, relies heavily on the family system and coping models to provide a foundation for services such as family support and financial counseling. Likewise, the chaplain implicitly recognizes the validity of these models in providing spiritual guidance.

The nurse may use a combination of the diagnostic, coping, and family system models as he or she provides clinical care and support to patients and their families.

The members of a transplantation selection committee, who must form an opinion on the likelihood that a potential candidate will comply with a strict medical regimen, cannot rely solely on one model. Their attempt to predict a patient's future behavior requires a complex understanding of the individual and his or her family and social milieu. Thus, the diagnostic, psychodynamic, and family system models must be employed together.

The health care policy-maker deciding whether to allocate financial resources to pay for transplantation needs a simple estimate of whether or not this expensive treatment really makes a difference in the life of most patients. The QOL model is a natural choice for this inquiry.

Alternative Models

Other psychological models exist, for example, "behavioral psychology," which explains human behavior as it is learned through reinforcement, and "existential psychology," which examines the meaning of life and death. Such models readily lend themselves to a better understanding of transplantation patients. Behavioral psychology, for example, helps us comprehend the problem of medication noncompliance as a learned behavior rather than as a moral failure. Existential psychology affords a framework for exploring the deepest concerns of patients as they struggle with life and death decisions.

Frontiers

What frontiers lie ahead in our attempts to better understand the psychology of the transplantation patient? One fruitful avenue of research may be found in the idea of "resilience," a concept borrowed from studies of childhood mental health (Rutter, 1987). These studies reveal that some children remain remarkably untroubled despite extraordinary levels of psychosocial stress. Exactly what psychological processes protect these individuals against the potentially deleterious stress of misfortune, abuse, or neglect is uncertain. The answers, however, may help us understand how some patients undergoing transplantation weather prolonged pain, disability, and repeated setbacks, yet emerge not only emotionally intact, but in some instances, even stronger than before. Ultimately, this understanding may aid in helping all patients who face the challenge of transplantation.

Sexual Performance Before and After Organ Transplantation

Ralph E. Tarter and David H. Van Thiel

Introduction

Since the introduction of antibiotics, combined with major advances in biochemistry, pathology, and preventative medicine, a substantial reduction in the prevalence of mortality from acute medical illnesses has occurred. In contrast, the number of deaths from chronic diseases has, annually, steadily increased since the beginning of the century. Many of these deaths are the result of self-inflicted injury from excessive alcohol and tobacco consumption or poor diet, but others are the consequence of the inability of modern medical technology to effectively treat the disease conditions. Hence, deaths from diabetes, vascular disease, liver disease, and pulmonary disease have not been sharply reduced in this century other than through modest advances made in the field of surgery. One significant advance in surgery is organ transplantation.

Organ transplantation stands at the vanguard of advanced medical/surgical technology for preventing the inevitability of death from certain chronic diseases. However, this technology itself poses a number of problems, particularly with respect to the current pressing issue of biological survival at the expense of a reduced quality-of-

From: *New Harvest* Ed.: C. D. Keyes
©1991 The Humana Press Inc., Clifton, NJ

life. Among the most important components of a satisfactory life-quality is sexual adjustment. The capacity to engage in intimacy and to achieve the benefits of rewarding sexual involvements is intrinsic to a healthy life in modern society.

Sexual performance is an important component, worthy of study with respect to organ transplantation. Its study must be viewed in the context of all other appetitive drives. Hence, sexual performance as a consummatory act is not to be considered in isolation from the other consummatory behaviors, specifically eating and drinking. Satisfactory sexual performance depends on a host of organic, psychological, and social processes. Attitudes and values that shape the motivation for such behavior influence the capacity to achieve satisfaction from sexual relationships.

The following discussion addresses the key issues pertinent to sexual functioning before and following organ transplantation. It should be pointed out that there is a paucity of empirical research investigating this aspect of life-quality. No systematic prospective study of sexual performance has been conducted on patients undergoing organ transplantation. Hence, much of the following discussion will reflect the authors' extensive experience with the organ transplant population and will consist of information gleaned from several hundred clinical interviews of patients who have undergone organ transplantation.

Sexual Performance Before Transplantation

Chronic disease is commonly associated with dysfunction in a variety of daily routine activities. This is caused by the direct effects of illness in reducing mobility and alertness, and from discomfort and pain consequent to the disease. In addition, psychological distress, reactive to the disease, negatively affects attitudes, motivation, and perception of functional capacities. For example, an individual with a fatal heart condition is limited with respect to capacity to exercise or perform manual tasks. Such an individual is also likely to have high dispositional anxiety and fear; engaging in sexual interaction because of the cardiovascular stress is thus characterized by avoidance and apprehension. On the other hand, individuals who have advanced diabetes and are candidates for pancreas transplantation also have advanced peripheral neuropathy. Therefore, they are not able to obtain sexual fulfillment because of organic dysfunction. Jaundice often accompanies the end-stage of liver disease. This affects both perceived attractiveness and physical well being, which, in turn, mitigates the opportunity for a fulfilling sexual experience. Thus, it can be seen that, depending on the particular illness for which the individual is a candidate for transplantation, there are many factors that could impinge on both the motivation and potential for satisfaction from sexual relationships.

One overriding consideration with respect to obtaining satisfaction from sexual intimacy concerns psychiatric and emotional status. There is strong empirical evidence demonstrating a high prevalence of psychiatric illnesses in candidates for organ transplantation. Symptoms of depression and anxiety are particularly common. These symptoms could adversely affect the capacity to attain sufficient sexual arousal as well as hold sexual interest. For example, high anxiety in the male is counterproductive for obtaining an erection. Depression and accompanying feelings of apathy, irritability, and agitation similarly will negatively affect the capacity to experience pleasure from sexual relationships. The emotional state of the individual is a major factor in determining whether sexual adjustment will be affected in individuals who are candidates for organ transplantation.

Factors Affecting Sexual Functioning Prior to Organ Transplantation

Disease Type

The specific chronic disease will have, as noted above, a significant effect on the individual's inclination to engage in sexual behavior. Of course, sexual behavior is not distinct from other aspects of daily living, nor is it to be considered outside of the context of other consummatory motivation. Yet, the chronic nature of the disease will influence the person's ability as well as attitudes regarding sexual motivation. Not only will it operate with respect to engaging in intimacy, but it may also affect the decision to procreate. For example, we have observed at the University of Pittsburgh that some patients with life-threatening liver disease will decide to have a child intentionally, even anticipating the fatal outcome of the disease. In contrast, other patients will intentionally avoid all forms of sexual contact so as to completely avoid the risk of complicating their already emotionally, socially, and economically fragile status.

Gender

Sexual experience, particularly the meanings attached to intimacy, differ greatly between the sexes. Physical attractiveness, for example, may be more important for a woman than a man, and, hence, disfigurement or changes in skin pigmentation are likely to have a more deleterious impact on the female sexual functioning. In contrast, in the traditional relationship, the male is often the initiator of sexual contact. If the disease poses risks from excessive exertion, as in heart disease, or produces a sharp reduction in energy, as in liver disease, the male's capacity to initiate sexual contact is reduced.

Age

Sexuality takes on a different meaning for the individual depending on age. Younger adults attach more importance to physical attractiveness and hedonistic experiences. Significantly, the short duration of the relationship among young couples is a factor in determining adjustment following onset of the disease. In long-term monogamous relationships, sexual interaction takes place in a different context, and, thus, increasing importance is attached to the values of intimacy, bonding, and nurturing, whereas less emphasis is placed on the hedonistic pleasures derived from the sexual act itself. Hence, depending on the patient's age, sexual adjustment prior to organ transplantation will be affected by the type of satisfaction that can be expected to be derived.

Dimensions of Sexual Performance and Adjustment Performance Factors

A number of factors influence the capacity to engage in sexual behavior. These factors should be considered as distinct from satisfaction or the hedonistic pleasure derived from sexual intercourse. Performance factors pertain to the capacity to initiate sexual interaction, attain sexual arousal, and the ability to sustain such arousal through completion of the sexual act. Where the disease condition has induced a state of apathy or fatigue from either reaction to psychiatric or medical treatments of the chronic disease, arousal mechanisms are attenuated and sexual performance will be impaired. The impairment can be variously manifest in the male as shown by an incapacity to attain or sustain an erection or to ejaculate after a reasonable period of stimulation. In the female, a common problem is difficulty in achieving significant levels of excitement so as to induce orgasm, along with discomfort due to incomplete vaginal lubrication and muscular tension. Failure to adequately consummate the sexual act further exacerbates the experienced psychological distress, thereby creating a spiraling effect on magnifying negative attitudes and aversion towards engagement in sexual activity.

Self-Concept

Self-esteem, if diminished by the illness, greatly influences one's capacity to establish intimate relationships. Self-concept can be reduced by diminished physical attractiveness. Additionally, self-concept is comprised of psychological components such as likability, sociability, and motivation. The extent to which individuals with chronic disease have impaired self-concept depends on whether there are rewards in the person's life that can adequately maintain ego integration. A negative self-image, therefore, will attenuate the individual's initiative to seek intimacy as well as reduce the motivation to respond to a partner who wishes to engage in intimacy.

Attitudes

Attitudes towards the disease as well as the partner are critical factors that determine sexual adjustment. Attitudes are shaped by both the physical and psychological changes occurring synergistically in the sick person and the partner. For example, a spouse or significant other who is confined to a wheelchair may convey attitudes that do not promote heterosexual contact. Similarly, a patient who is irritable, sleeps excessively or fitfully, or is frequently apprehensive or forgetful also does not promote motivation in self or the partner to engage in a sexual relationship.

Attitudes that are inculcated in the partner become major determinants of sexual motivation in the patient.

Sexual satisfaction—the pleasure derived from intimacy—not only depends on the capacity to adequately consummate the sexual act, but, additionally, depends on a host of psychological factors operating during the period of intimacy. Under conditions where performance is marginal, satisfaction may be very good if the partner accepts the functional limitations. On the other hand, satisfaction will be very low if the partner does not accept the limitations associated with the patient's illness. Thus, there is a dynamic relationship between performance and satisfaction that varies during the course of the illness and depends to a great extent on the quality of the spousal relationship. In this regard, satisfaction in the bedroom parallels satisfaction in the overall dyadic relationship. It bears on the issue of the extent to which the patient can be pleased and the extent to which the healthy partner is willing to please and accept pleasure in an intimate relationship.

In addition to the motivation to bring pleasure to the partner, there is the common problem of harboring attitudes and fears that could be either justifiably or irrationally held. Under such circumstances, sexual intercourse may be perceived as dangerous and as presenting a burden to the patient and/or partner. In this regard, the psychological characteristics of the healthy partner cannot be overemphasized as a major determinant of whether or not performance and satisfaction in sexual intercourse is affected by the disease and to what extent this has occurred.

Pertinent Findings Relevant to Sexual Adjustment in Patients Prior to Organ Transplantation

Despite the paucity of systematic empirical investigation of this issue, the available evidence indicates that sexual adjustment and performance is significantly disrupted in a substantial majority of patients prior to organ transplantation. This is not only caused by the specific effects of the disease on reproductive organ functioning, but also relates to the overall psychological context of having the chronic and inevitably fatal disease. There is no particular sexual activity pattern or level of sexual adjustment prior to organ transplantation that characterizes the population, but rather the confluence of variables described above interacts in a highly individualistic fashion to determine both performance capacity and quality of satisfaction for each patient and their partner. Moreover, along the dimension of severity, it appears that sexual performance and adjustment may range all the way from marginal or no impairment to substantial disruption of quality-of-life. Hence, no general conclusions can be drawn with respect to sexual functional status prior to organ transplantation for the population of patients undergoing this procedure. It is also important to stress that sexual performance should be viewed within the total context of the individual's overall life-style, the meaning attached to sexual interaction, and the extent to which the marital relationship has emphasized the importance of sexuality. The more important sexuality is in the relationship, the more likely that even a mild disruption will have an adverse impact on the quality of the dyadic relationship. Up to 80% of patients who are candidates for transplantation experience some decrement in sexual drive, performance, or satisfaction at some time during the course of their illnesses. This finding underscores the prevalence and importance of sexual dysfunction.

Further complicating the process of sexual adjustment is the fact that there is widespread ignorance regarding sexual performance and adjustment. Most of the population are ignorant of reproductive physiology and are simultaneously lacking in knowledge about their disease. This state of ignorance is often not compensated for by education from either the attending physician or social worker and, hence, the chronically diseased patient is likely to harbor unwarranted and even irrational conceptions about their sexuality and ultimately their personal identity.

Sexual Functioning
Following Organ Transplantation

There is little empirical research specifically investigating sexual performance and satisfaction following organ transplantation. Theoretically, the extirpation of the diseased organ could restore the person to normal physical and psychological status. This, however, does not typically occur for a variety of reasons. First, the patient is often maintained on medications, especially steroids, that could impact adversely on appearance, mood and behavior, and, consequently, sexual performance. Also, the effects of lifelong immunosuppressant medication on sexuality has not been investigated, but the fact that one must always be concerned with impending infection can be expected to reduce the inclination to pursue healthy sexual involvements when there is physiological stress and exposure to infection. Second, organ transplantation does not immediately, and perhaps may not ever, eliminate all aspects of the chronic effects of the disease for which the organ transplantation was originally performed. For example, peripheral neuropathy in diabetic patients may not be resolved entirely following pancreatic transplantation. This will have a direct physical effect on the capacity to engage in sexual performance because of reduced sexual potency, especially in men. Third, patients undergoing organ transplantation often experience new life stresses that could interfere with sexual adjustment and performance. Economic hardship, new strains in the marriage, difficulty in resuming employment, and, often, the presence of coexisting psychiatric and emotional disturbance are complicating factors for many individuals following organ transplantation.

Another factor to be considered relates to the uncertain outcome of organ transplantation. Success is not guaranteed. The quality of physical health and psychosocial adjustment is highly variable. The reasons for the variability of outcome are not entirely known but probably relate to the chronicity and type of disease, intraoperative procedures, susceptibility to infection, and complications arising from the organ transplantation procedure. All of these factors will, of course, ultimately influence the quality of sexual adjustment as well as overall quality-of-life.

Reconciling Health Status and Psychosocial Adjustment

Health status and psychosocial adjustment can be easily disrupted. For example, a sudden change in health status commonly produces a change in psychosocial adjustment. Conversely, a major change in lifestyle can induce changes in health status. Sexuality, encompassing both the dimensions of performance capacity and satisfaction is, thus, the product of the synergistic and reciprocally dynamic relationship between health status and psychosocial factors. A disruption in either is likely to affect the other of which sexuality is but one component. For these reasons, it is important that sexuality be viewed as an integrated component of an overall life-style rather than seen as a unique or specific topic of interest. Hence, sexuality cannot be viewed simply within the context of ideals, particularly as espoused in the media or emanating from preconceived notions or fantasy, but rather, sexual performance and satisfaction should be calibrated to the individual's perception of what is reasonable for them given their medical condition, values, and life situation.

Intervention Strategies

Within the past two decades, major strides have been made with regard to treating sexual problems that have both organic and psychological causes. Indeed, sex therapy is a recognized professional specialty, and, in many states, practitioners are licensed or certified. It is important that the physician attending the patient undergoing organ transplantation and monitoring progress following this surgical procedure be aware of the importance of sexuality as an integral component of life-quality. Where disturbances are noted, referral to specialists in sex therapy would obviously be advised. This depends on a patient/physician relationship that is characterized by candor and an understanding by both the patient and doctor of the importance of psychosocial factors as determinants of life-quality following surgery.

Summary

Organ transplantation has been shown to improve survivability for a variety of chronic disease conditions where extirpation of the diseased organ can be replaced by a normal functioning organ. Chronic disease affects multiple spheres of functioning, including sexuality. Not surprisingly, both sexual performance and satisfaction are disrupted in a large majority of patients with chronic disease, particularly those where the disease has fatal consequences. No single profile emerges regarding sexual performance or satisfaction in patients having chronic disease, but rather the particular manifestations of sexual dysfunction depend on a variety of medical and psychosocial parameters outlined in this discussion. Similarly, the same general conclusion can be drawn regarding sexual adjustment and performance following organ transplantation. One conclusion that can be drawn, which appears to have consensus among the professional community, is that sexuality is only one component of complex processes comprising life-quality. The degree to which sexual performance and satisfaction are disrupted in patients before or after transplantation surgery is consistent with disruptions in other spheres of life functioning. Currently, there are standardized assessment procedures for the objective and comprehensive evaluation of sexuality. Interventions for sexual-related problems have also been developed. There is, thus, every reason to believe that these problems, where manifest, can be effectively managed. However, because the medical community has historically focused primarily on the medical sequelae of transplantation, education is required to expand their understanding and appreciation of psychosocial processes.

Body and Self-Identity

C. Don Keyes

Introduction

Transplantation causes a crisis in our relationship to the body. This crisis involves more than what is done to bodies; it affects what they symbolize, since the new surgery challenges existing attitudes towards the body and confronts us with unexpected and unusual ways of interpreting what bodies fundamentally are. The one most shocking discovery is that organs, tissues, and genetic material are no longer the fixed properties (possessions?) of a particular individual. Instead, they are increasingly interchangeable and, as Winslade and Ross (1986) suggest, this produces a crisis of self-identity:

> Transplants on the grand scale require us to think about who and what we are. A cornea here and there; a pint of blood at a bad time—these are not enough to create concern about personal identity. But when we begin to make promises to people who need and receive hearts and lungs and kidneys, or kidneys and livers, or more, then we need to think about ourselves in a way that will make sense of these substitutions.

Transplantation threatens self-identity in order to extend life, but the crisis it causes goes beyond revealing the flexibility of the body's composition. It also confronts us with contradictory concepts about what a person is. On the one hand, we are embodied brain-mind unities, and our organs symbolically continue to partake of that unity. On the other hand, a person is mind because life is defined in terms of brain function, and body parts are recycled from

From: *New Harvest* Ed.: C. D. Keyes
©1991 The Humana Press Inc., Clifton, NJ

one core of cerebral activity to another, as the quotation by Winslade and Ross in the first chapter suggests. Uncertainty about whether a person is a brain-mind unity or essentially mind alone springs from the still more basic question of how the brain and the self-identity are related. When brain death occurs, a self is no longer present even though heart beat and respiration can be sustained artificially.

The interchangeability of body parts also leads to uncertainty about what kind of objects they are. Although they are not absolutely fixed properties, body parts are not ordinary things either, like interchangeable machine parts. The fact that body parts were previously attached to a core of cerebral activity means that they were once parts of a whole that lived, loved, and was loved as a whole. Some symbolic determination of that whole continues to live with the life of the organ. This makes body parts represent something more than mere objects to prospective donors, to their families, and to society. Similarly, body parts are more than mere objects to the recipient, since they extend the recipient's life and represent the temporary defeat of death. The prospective donor sometimes views the gift of his organ as a continuation of an element of his own life after he dies. The recipient may see the new organ in the same way and as a result have feelings either of gratitude or guilt, even if such feelings have no rational foundation. Thus, body parts are objects laden with meaning because they symbolize the self, even when separate from it.

The crisis of self-identity caused by transplantation takes two forms, precipitated by two kinds of uncertainty: the "Crisis of Composition" and the "Crisis of Brain and Self."

Crisis of Composition

As has been shown, transplant surgery challenges the conventional view that body parts are the fixed properties of the selves that own them. Organs are no longer fixed properties because medical technology now makes them increasingly interchangeable. The belief that the family owns one's body after one dies is now challenged on moral grounds since lives can be saved if organs can be recycled and refusal to do so will cause others to die earlier. The new surgery overthrows ordinary ways of looking at the body's composition. This has produced a new kind of crisis with a distinctive type of symbolism.

An organ recipient sometimes experiences a changed body image, such as pain, itching, jaundice, surgical scars, changes in body size, and the like, which is caused partly by the effects of the recipient's disease or recovery. Emotional trauma of illness, surgery, and drugs can also contribute to changed body image, and so can deprivation of sex and alterations in social relationships. As Kern explains (*see* "Psychological Perspectives on the Process of Organ Transplantation," this volume), organ recipients normally adjust to the graft without undue emotional disturbance. He also mentions that unusual emotional states, such as those emphasized by the earlier psychiatric literature of organ transplantation, can occur. Whereas phenomena like those described in the following account of some of that literature are not always present, they reveal a great deal about a way body parts can take on symbolic meaning. In these extreme cases, the changed body image seems to be due to the fact that a foreign organ has been added to a body, as Castelnouvo-Tedesco claims (1978, 1981). He also contrasts (1973) conventional (life-saving) operations and transplant (life-extending) operations and claims that the latter causes an enlargement of the patient's body image:

> Moreover, life saving operations, by removing a diseased part, appear to restrict or limit the body image...By contrast, life-extending operations enlarge the body image...the individual must make room for his additions so that something which previously was nonego may come to be felt as part of the ego.

Body image is "a fluid structure," but "it takes time to adapt to changes in the body's architecture," and even gradual and expected changes such as aging cause considerable anxiety. Sudden changes in the body's composition, such as restriction in life-saving surgery or enlargement of it in life-extending surgery, can cause an even greater crisis in body image. The new body part is more than an organ; it is also a symbol, according to Castelnouvo-Tedesco (1981):

> The new organ is not psychologically inert. Rather it has a psychological meaning and activity which it displays from the very beginning. The realization that the organ comes from another person has very serious implications...The transplant, then, is not just a piece of plumbing but a symbolic representative of another human being and of the relationship, both real and fantasied, to that person.

Recipients often struggle with the extraordinarily powerful symbolism of what has been added to their bodies before they accept it. Castelnouvo-Tedesco (1973) claims that such patients experience "[t]he new acquisition as an introject, is an object or part-object that immediately becomes affectively very active." He points out that this often produces ambivalence: "[O]n the one hand, the euphoria associated with the possession of a new life-giving and strength-enhancing source; on the other hand, paranoid dread and panic arising from the presence within of an object which now has become a dangerous persecutor." Not all patients experience this ambivalence. Some emotionally adapt to their new organs from the start. Others, however, deny the ambivalence, whereas many work through the effects of the symbolism of the body's enlargement by externalizing the organ or by regressively identifying with the donor of the organ, or by some combination of the four.

Struggle to Adapt

Denial is typical or, according to Castelnouvo-Tedesco (1981), "massive," making it "difficult for the patient to consider the broader aspects of his life situation." Patients often give little thought to their progress or future plans, but are restricted to the here and now, a fact that does not necessarily signify denial. Castelnouvo-Tedesco (1981) wisely points out that therapeutic help given to such patients ought to aim at support rather than increasing insight. More recently, Mai (1986) studied 20 heart recipients and found that 18 engaged in denial:

> The most striking finding was the presence of denial in 18 of the 20 patients (90%). Denial was expressed towards the graft, the donor, or both.
>
> Graft denial (seven subjects) was expressed in comments such as 'I have no thoughts about my new heart,' 'I've no feelings about having a new heart,' 'I try to forget about the new heart,' and 'I am not bothered about having another heart.'
>
> Donor denial (five subjects) was expressed in remarks like 'I have not asked where it (the heart) came from,' 'I never think about the donor,'...In addition six subjects showed denial towards both graft and donor.

Mai explains that this kind of denial legitimately functions to protect the recipient and help him to adapt:

> An essential requirement for adaptation is the need to incorporate a new body image of a vital organ. This organ, in turn, was harvested from a dead donor, frequently anonymous, who was the victim of an unfortunate disease, injury, or accident...The results suggest that posttransplant adaptation may indeed be promoted if the patient is given a time for silence.

Externalization of the graft is one way the symbolism of an enlarged body image is expressed. Some patients treat their new organ as if it were either entirely or partly a foreign object. Muslin (1971), who studied kidney recipients, claims that some patients identify the organ with a new body or fetus, "a reaction found in both sexes, with the patient identifying himself with the protective, caretaking mother." One of his male patients said that "[t]he kidney has to be nursed like a baby, you feel like a mother to it." Muslin writes that patients often treat the graft as a foreign body and say, for example, that it "feels funny" or "sticks out" as if someone "were to have a new eye inserted." Mai claims that some patients think of their new heart as an appendage, treat it as if it had a life of its own, or speak of it in the third person. One patient, for example, who received a kidney from her sister said, "[s]he's doing pretty good" when she talked about her kidney. Symbolizing the graft as external might be a means of denial, Basch (1973) points out, but it can also accompany favorable psychological adaptation, as it showed with the patient just quoted "who viewed the kidney as still belonging to her sister but enjoyed an uneventful postoperative course psychologically and continued compatible relations with her sister."

Identification with the donor by the recipient is not always as adaptive as the case mentioned in the last quotation. More often, it is a painful, even regressive, struggle, since the symbolic power of the new organ can cause a "psychological fusion of the ego boundaries of donor and recipient" (Basch, 1973). The limits of self-identity expand to the donor, whether he is living or dead. Castelnouvo-Tedesco (1981) claims that the recipient sometimes acts as if he had internalized the donor:

> In short, one finds that the transplant is not just a body part, but that psychologically it represents the whole person...The process is somewhat as follows. After transplantation, the transplanted organ promptly achieves mental representation, i.e. it registers in the mind both as an anatomic part and as a symbolic representative of the donor. It also relates to existing introjects, especially those of the early parental figures. It unites with these and activates them.

Researchers often report two types of identification. First, there are incorporation fantasies produced by the belief that some personal traits of the donor have been added to the self through the graft. These traits might be real or imagined and their incorporation either welcomed or dreaded. Such fantasies are normally less disturbing if the donor is of the same sex and age group as the recipient. Patients sometimes feel that they gain new strength, masculinity, or femininity, depending on who the donor was, or they imagine the graft incorporates undesirable traits or threaten their sexual identity. Such fantasies could include both tangible and intangible qualities, as Castelnouvo-Tedesco (1981) reports, "generosity or altruism, artistic talent, religiosity, helpfulness, aggressiveness, capacity to speak a foreign language, physical size (e.g., slimness)." Lunde (1969) reports:

Some patients have felt that by receiving the heart of another person they might take on some of the personality characteristics of the donor. One man literally decided that the day of his transplant was his new birthday, which he planned to celebrate from then on. He felt that he had been born again and was 20 years old. This was a 42-year-old man who had received the heart of a 20-year-old.

Even if the anonymity of the donor is safeguarded, as is usually the case with cadavers, transplant recipients often imagine that the donor had traits that they fantasize will become a part of themselves. Fantasies about acquiring such traits appear to be greater with heart transplants, since livers and kidneys seem not to be laden with as much distinctive symbolism as hearts. There is evidence that liver recipients tend to experience fewer incorporation fantasies than other transplant patients.

Second, ambivalence about the right to have the organ often torments patients. On the one hand, a recipient is thankful to the donor (or family) for the gift of life. Expressing thanks is a problem because there is no adequate way of repaying this kind of gift. As Fox (1978) claims, the recipient

...has accepted an inherently unreciprocal gift from the donor...The recipient can never totally repay the donor for his priceless gift. As a consequence, the giver, the receiver, and their families may find themselves perpetually locked in a creditor-debtor vise that constricts their autonomy and their ability to reach out to others.

The nonreciprocal nature of the donor/recipient relationship frustrates the recipient's overwhelming gratitude toward the donor even though the recipient has good will toward the donor. On the other hand, the recipient often feels guilty about having the organ and might be obsessed with the thought that he caused the donor's death or at least willed it. Castelnouvo-Tedesco (1973) states that at some level, the patient views the transplant

...as something that does not belong to him and to which he has no rightful claim. During regressed mental states, guilt about having stolen the organ may occur together with the feeling that his own essential characteristics have been altered as a result of possessing inside a part from another human being.

Muslin's kidney recipients showed mixed reactions of elation and excitement "combined with or (at times) replaced by concern about the donor's health" or reacted in an equally common way by denying that anything "special has been given to them by the relative or cadaver." Although each patient's way of being elated, excited, or concerned was unique, Muslin notes some typical ways of struggling with the problem. Some defended themselves against guilt and shame by denying, avoiding, or repressing their recognition of the gift: "My brother—oh, he is fine, he's healthy, he's a football player. He doesn't need two kidneys." Others experienced the guilt and shame of feeling that the gift destroyed, limited, or curtailed the donor's life: "I feel like I'd be taking his life if something happened [to his other kidney], and this I don't think I could live with." Such diverse reactions as "intense concern for the other person's life, sudden feelings of being identified with the person, and feelings of justification in defense of having taken the kidney" typically accompany restitution measures. Sometimes, restitution takes the shape of "Siamese-twin" type identification in which the recipient might feel the organ is sick if the donor is sick. At other times, restitution turns into the opposite of this when recipients justify having the organ and claim that the donor did not need it. Externalization of the guilt occurs when patients experience fear of punishment and guilt in ways that are often archaic and regressive. Muslin's (1971) example of a kidney recipient who believed his graft was about to die illustrates this.

> He...felt the presence of a ghostlike spectre in hypnogagic state after his transplant. One day he arose in an anxiety attack because he had been in a dream-like state of hallucinosis, with the experience of feeling the presence of a ghost connected with his cadaver-donor, which was threatening him.

Adaptation is emotional integration of the graft. This process is a symbolical change in the way the patient thinks and acts with respect to his body. Muslin describes the psychological process of integration as an analogy of physical acceptance of the organ. According to him, "The assimilation of the new tissues has as a counterpart the ego's integration of the new object in its core (at first periphery)—a psychological transplant." This typically occurs after two previous stages. The first, "foreign body" stage, is the externalization mentioned above, during which the graft is still separate and alien to the ego. After this, but prior to adaptation, the second stage or "partial incorporation," occurs, during which patients talk less about their transplants and appear not as uneasy about them, not as in awe of them, or not as interested in them as before. Complete incorporation occurs, according to Muslin (1971), when

...the patient reports an acceptance of the new organ to the point of not being aware of it unless questioned or unless a procedure is to be performed on the organ. Psychologically it would seem to indicate that the image of the organ is more completely integrated with the internal images of the patient's body and ego.

In other words, adaptation or complete incorporation means that the new organ has become "part of the psychological apparatus, finally becoming merged with the individual's self-representations." Muslin and other specialists recognize that adaptation is not necessarily a fixed condition, since the patient's struggle with death and conflicts associated with what Castelnouvo-Tedesco (1973) calls the "introject" invites regression to what is fantastic, primitive, and instinct-ridden. Regression occurs especially when biopsies or even more routine procedures must be performed.

The Body and Its Parts

The symbolism of enlargement, which lies at the base of much of the patient's emotional struggle to adapt to a new organ, dramatically reveals the bodily side of self-identity, as does some of the other symbolism of bodily parts reported by the early psychiatric literature mentioned above. Resistance to internalizing the organ psychologically, ambivalence towards the donor, incorporation fantasies, and other elements of the struggle to adapt point to belief in the corporeal nature of the self. Body parts are life-saving objects that also symbolize their former self. Patients often relate to the grafted organ as if the personality of the donor were, in some sense, still attached to it. As a result, the patient acts as if real or imagined attributes of the donor's self were incorporated along with the organ. Also, the donor, or his family, sometimes wants a part of his bodily identity to survive his death and live in and through a recipient. Viewed in one way, these animistic bonds between recipients and donors are regressive or subrational. Yet viewed in another way, the same symbolism that binds donor and recipient also expresses a truth, even when that truth is experienced in ways that are bizarre and regressive. The recipient often interprets the graft as symbolizing the donor because a body part does in fact differ from a machine part. Even though it is no longer the donor's organ, it once was and that makes the organ different from any other kind of object.

Crisis of Brain and Self

Life-extending surgery also confronts us with the question of how the brain relates to the self, to return to a concern expressed in the second quotation by Winslade and Ross at the beginning of this chapter. Organ grafting, as it exists today, initially seems to point in two irreconcilably different directions: dualism and monism. The first theory claims that mind and brain are two different realities that influence each other, whereas the second holds that they are the same reality.

Dualistic Tendency

Modern medical science would seem at first glance to be inclined to monism (a physical monism in fact) rather than dualism. It does not deal with souls, but with brains. We live while our brains live; we die when they die. This very focus on the brain, though, may imply a new and unintended kind of dualism. There seems to be an implicit assumption that as long as the brain is intact, the self remains intact. Other body parts are interchangeable and transplantable, but we have already seen that self-identity is also connected with other organs. What of the rest of our bodies? Are we not also our faces, our glands, our metabolism, our stature (tall or short), our muscular coordination (good or bad), our allergies? It is true that we cannot exist without a functioning brainstem and that the "higher" operations of consciousness depend on a functioning neocortex. It is not true that we are our brains, somehow distinguishable from the rest of our bodies. Jonas (1980) has called attention to this new dualistic tendency:

> I see lurking behind the proposed definition of death, apart from its obvious pragmatic motivation, a curious remnant of the old soul-body dualism. Its new apparition is the dualism of brain and body. In a certain analogy to the former it holds that the true human person rests in (or is represented by) the brain, of which the rest of the body is a mere subservient tool. Thus, when the brain dies, it is as when the soul departed: what is left are 'mortal remains.' Now nobody will deny that the cerebral aspect is decisive for the human quality of the life of the organism that is man's...But it is no less an exaggeration of the cerebral aspect as it was of the conscious soul, to deny the extracerebral body its essential share in the identity of the person. The body is as uniquely the body of this brain and no other, as the brain is uniquely the brain of this body and no other...My identity is the identity of the whole organism, even if the higher functions of personhood are seated in the brain. How else could a man love a woman and not merely her brains?...Therefore, the body of comatose, so long as—even with the help of art—it still breathes, pulses, and functions otherwise, must still be considered a residual continuance of the subject that loved and was loved.

Monistic Tendency

The symbolism of body parts resists dualism by announcing the bodily nature of the self, when, for example, a recipient interprets a graft as if attributes of the donor were permanently attached to it. Whereas some recipients experience this superstitiously and regressively, the underlying symbol suggests that the self is essentially corporeal. This symbol represents body parts as components of their former self, residual traces of a brain-mind unity that they once embodied. Such body parts symbolize the donor's personality in and through their materiality, not apart from it. The monistic tendency of naive body symbolism was implied throughout the discussion of the struggle to adapt, but there are at least two other ways it can be expressed.

First, emotional resistance to brain death is more than merely a survival of prescientific thinking. It might be based on the tendency to represent body parts as symbols that still point to the self with which they were once unified. As long as the brain functions, I am my body, and its parts acquire a surplus of meaning that is not totally extinguished for others when my brain ceases to function. A cadaver, whether its heart beats or not, is also a dead body. It is a collection of parts that were once attached to a functioning brain, thus making it and organs taken from it differ from ordinary things. As Jonas (1974) also writes,

> The body is uniquely the body of this brain and no other, as the brain is uniquely the brain of this body and no other. What is under the brain's central control, the bodily total, is as individual, is as much "myself," as singular to my identity (fingerprints!), as noninterchangeable, as the controlling (and reciprocally controlled) brain itself. My identity is the identity of the whole organism, even if the higher functions of personhood are seated in the brain...How else could we lose ourselves in the aspect of a face? Be touched by the delicacy of a frame? It s this person's and no one else's. Therefore, the body of the comatose, so long as—even with the help of art—it still breathes, pulses, and functions otherwise, must still be considered a residual continuance of the subject that loved and was loved, and as such is still entitled to some of the sacrosanctity accorded to such a subject by the laws of God and men. That sacrosanctity decrees that it must not be used as a mere means.

Although they avoid using the exact term, Youngner et al. (1985) point to potential ethical conflicts in the "harvesting" of organs. On the one hand, the well being of the recipient whose life is preserved through the procedure takes precedence over all other goals. On the other hand, the removal of organs seems to show disrespect to the dead and, in some sense, violates the person of the deceased. This suggests that neither the cadaver, which is dead even though it might not appear to be so, nor the retrieved organs, which are living, are mere things. Youngner et al. describe the removal of organs from the standpoint of its emotional effect on the operating room staff who normally "accept brain death on an intellectual level." The procedure deviates from normal operating room practice since, in this case, the dead are sent into surgery even though they do not always appear to be dead; they might be under general anesthetic like living patients:

> However, the most difficult deviation from normal routine occurs when it is determined that the life functions of the dead patient need not be supported any longer, because the organs that required perfusion have been removed. At this point, after long hours of arduous surgery, instead of discontinuing the anesthesia and waking the patient, the anesthesiologist must turn off the ventilator, thus halting all the remaining life functions; he or she then leaves the head of the operating table and, often, the room. The surgeons who have not left with the viable organs close the body cavities—generally in one pass, using coarse retention sutures and large needles. In some instances, surgeons remain to retrieve useful organs, such as eyes or skin, that do not require constant perfusion. At the end, one or two nurses are left alone to tidy up the body and prepare it for transport—not to the customary recovery room or intensive care unit, but to the morgue.

> There are other important deviations. The objective of surgery is usually the preservation of healthy tissue, not its removal, and the welfare of the patient on the table, not someone else. (Even when organs are taken from living donors—e.g., a kidney—the health and safety of the donor remain primary concerns.) Operating room personnel may be shocked by the mutilating nature of some procedures, such as the removal of skin or long bones. Organs such as

the heart may have special symbolic meaning. Indeed, dead bodies are not ordinarily treated in such a fashion, except in specially designated places, such as the autopsy table or medical school dissecting room, which are far removed from treatment settings.

Second, persistent ethical objections to keeping a brain dead body "alive" might be due to the surplus of meaning that distinguishes a dead body and its parts from mere things, as explained above, but especially when heart beat is artificially sustained. There would be a practical, perhaps even a commercial, advantage in not turning the respirator off after the brain is dead. Jonas (1974) writes,

> I have no doubts that methods exist or can be perfected which allow the natural powers for the healing of surgical wounds by new tissue growth to stay alive in such a body. Tempting also is the idea of a self-replenishing blood bank. And that is not all. Let us not forget research. Why shouldn't the most wonderful surgical and grafting experiments be conducted on the complaisant subject-nonsubject, with no limits set to daring? Why not immunological explorations, infection with diseases old and new, trying out of drugs? We have the active cooperation of a functional organism declared to be dead: we have, that is, the advantages of the living donor without the disadvantages imposed by his rights and interests (for a corpse has none). What a boon for medical instruction, for anatomical and physiological demonstration and practicing on so much better material than the inert cadavers otherwise serving in the dissection room! What a chance for the apprentice to learn in vivo, as it were, how to amputate a leg, without his mistakes mattering! And so on, into the wide open field. After all, what is advocated is the full utilization of modern means to maximize the value of cadaver organs. Well, this is it.

Gaylin (1974) has a similar concern:

> These cadavers would have the legal status of the dead with none of the qualities one now associates with death. They

would be warm, respiring, pulsating, evacuating, and excreting bodies requiring nursing, dietary, and general grooming attention—and could probably be maintained so for a period of years. If we chose to, we could, with the technology already at hand, legally avail ourselves of these new cadavers to serve science and mankind in dramatically useful ways. The autopsy, that most respectable of medical traditions, that last gift of the dying person to the living future, could be extended in principle beyond our current recognition. To save lives and relieve suffering—traditional motives for violating tradition—we could develop hospitals (an inappropriate word because it suggests the presence of living human beings), banks, or farms of cadavers which require feeding and maintenance, in order to be harvested. To the uninitiated the 'new cadavers' in their rows of respirators would seem indistinguishable from comatose patients now residing in wards of chronic neurological hospitals.

Objections like those expressed by Jonas and Youngner et al. are not failures to accept the truth of brain death. Instead, they express respect for the life that bodies and body parts continue to represent in an at least residual way after brain death. Even though the difference between life and death should be decided in terms of brain death alone, the relation between the brain's activity and the humanness it grounds is not purely mechanical and brittle. The symbolism of the body and its parts revealed in the "Crisis of Composition" and concerns like those expressed by Jonas and Youngner et al., point to what could be called an unreflective, lived monism. Recognizing the compatibility between that and the theoretical monism based on neurobiological evidence puts the "Crisis of Brain and Self" into perspective. In this way, transplantation's first crisis of body and self-identity sheds light on the second. Whereas dualism clearly dominates the popular consciousness of the brain-mind relation, there are also monistic tendencies in public attitudes towards bodies and their parts.

Brain death and the interchangeability of organs from one brain-mind unity to another are contrary to conventional interpretations of the self, but they are not contrary to scientific reason, and altering the body's composition for legitimate reasons can be defended ethically. Brain function is the biological basis of personal individuality, namely our awareness of pleasure and pain, intellect, decision, and feelings previously attributed to the heart. At the same time, when brain function ceases, a surplus of symbolism attaches itself to the body parts and they preserve emotional traces of their former identity. This attests both to the material nature of a person and to the symbolic meanings that make bodies different from all other kinds of matter.

PART FIVE

Religious Perspectives

Introduction
to Religious Perspectives

Walter E. Wiest

Introduction

Religious symbols lead to theological interpretations of transplanation. These demand consideration since they will inevitably play a role in social ethics and in public debate about policies to be put into effect. Religion plays a significant part in our society and religious ethics have already contributed to contemporary medical ethics. The next two chapters consider Judaism and Christianity in detail because they are the most dominant and influential expressions of religion in western culture. To redress the balance somewhat, this introduction gives an all too brief review of medical ethics in three other world religions, all of them having some representation in the US today. This review is especially indebted to Veatch (1981).

Islam

The most basic item of Islamic belief is that Allah is omnipotent and that everything is subject to his will. Hence, any human action that seems to be too presumptuous, that appears to go too far in trying to take control of things out of God's hands, may be condemned. In fact, anatomical dissection, organ transplantation, and birth control have all raised objections from some Moslems.

From: *New Harvest* Ed.: C. D. Keyes
©1991 The Humana Press Inc., Clifton, NJ

On the other hand, this same item of belief has engendered a prohibition against killing, at least in some forms (e.g., infanticide). Life is worthy of respect because it is God's gift. In the Koran, it is stated that "whosoever killeth a human being for other than manslaughter or corruption in the earth, it shall be as if he had killeth all mankind, and whoso saveth the life of one, it shall be as if he had saved the life of all mankind" (Surah V: Ayeh 32). Therefore, there is an obligation to save life, and this supports medical treatment and the development of medical art and science.

In Islam, as in Christianity, belief in resurrection of the body has raised objections to damaging bodies by taking organs from them for transplant purposes. How can a body be resurrected if it has lost some essential parts? In response, some Islamic theologians claim that God, who creates out of nothing, is capable of recreating bodies. Of course, this does not justify careless or irresponsible treatment of bodies. We are the stewards, not the owners, of our bodies. Islam is similiar to Judaism and Christianity in seeing the body as part of God's good creation. Even the highest regard for spiritual values is not permitted to detract from a responsibility for physical health and well being or from the value of medical knowledge and skills that can preserve and improve them. In fact, the traditionally Christian West owes an incalculable debt to Moslem medicine from which it learned so much in the medieval and early modern eras.

All this makes room for approval of donation of bodily parts for transplant operations, provided, of course, that there is due consideration of risks and benefits. Sachedina (1988), an Islamic theologian, says this:

> It is possible to summarize Islamic views on organ transplant by pointing out the underlying principle of saving human life. Whereas the limited right of a person is recognized on his/her body, which is a trust from God and as such is to be preserved and respected, donation of organs from both living and dead has been regarded as permissible

in the jurisprudence. However, the prerequisite in the case of a living person is that his/her life is not endangered, whereas in the case of a dead person there must exist his last will of testament permitting thus, or the permission of his relatives.

Hinduism

Westerners have often held a view of Hinduism that could cause us to wonder whether it offers any basis for a medical ethic. Embodied life is seen as one of sorrow and suffering from which we seek release (*moksha*). We are bound to the "wheel of *samsara*", fated to experience repeated incarnations in which we must serve out our sentences for offenses committed in previous existences (*karma*). Final salvation, if we can achieve it, is *nirvana,* a state of being that both transcends and negates the conditions of physical existence. The disciplines prescribed for salvation seem to include ways of punishing the body for the good of the soul. How can such a view take medicine seriously?

The truth is that Hinduism is much more complex and subtle than this. For instance, in ancient Hindu literature, there is a whole corpus of material dealing with medical ethics (the *Ayurveda*). A religious tradition that took the ills of the body to be insignificant could hardly have produced such a corpus. There would have been no medicine to produce ethical questions about how it should be practiced.

Hinduism features a profound respect or reverence for life, both human and nonhuman, as well as a concern to relieve suffering. "Life," here, means physical, embodied life, as is clear from the traditional injunction against killing, either humans or other animals. The principle of nonviolence (*ahimsa*) with which modern Westerners have some acquaintance, especially through the impression made by Mahatma Ghandi, reinforces this injunction. Taken together, these two items of belief support the ancient counsel that was given to Hindu medical students in an initiation ceremony at the start of their training. Veatch explains how this required them to make every effort to relieve their patients and not abandon or harm them for any reason whatsoever.

Hinduism and Hindu medical ethics have changed over the centuries, partly due to influences from Buddhism and Islam and, more recently, from the West. Hindu physicians are likely to act these days on some combination of traditional Hindu ethics and modern Western ones. Like their counterparts of Christian or Jewish faith, they do not always follow the official teachings of their religion. Yet some elements of Hindu belief are still influential.

In regard to transplants, Hinduism seems not to present the problem that belief in resurrection of the body has raised (in the past, anyhow) for the harvesting of transplantable body parts. Reincarnation is not resurrection. Although the body is important in one's present life, one's essential self is the soul that is carried over from one embodiment to another. This would seem to allow for approval of the use of cadavers for organ donation.

One Hindu belief might create difficulties; the doctrine of *ahimsa*. Of course, when it is taken as a counsel to do no violence or injury, as Veatch suggests, it sounds much like that basic principle of Western medicine, "above all, do no harm" *(primum non nocere)*. But, apparently, it can be taken to mean nonintervention in a sense that might lead a physician, as Veatch also notes, to limit treatment to relieving suffering while refraining from further treatment that might be more positively beneficial. In such cases, medical interventions like transplants might seem unacceptable.

Given the complexity of Hindu thought, however, and the mix of ideas that history has brought to Hinduism, there seems no need to assume that a Hindu would necessarily be opposed to transplantation, and might well be supportive of it.

Buddhism

Since Buddhism sprang from Hinduism and shares much of the same attitude toward the spiritual and the physical, soul and body, it has been subject to similar misunderstandings by Westerners. In truth, there is a long history of Buddhist contributions to medicine and medical ethics.

Buddhism is primarily devoted to the relief of suffering. Admittedly, its concern is primarily with the damage done to the soul by "desire" (*tanha*), i.e., by the wrong kind of attachment to material or worldly things that promise happiness but actually deceive us and let us down. Yet, like Hinduism, it includes a reverence for life. One of its rules of conduct is "Do not kill" (a part of "right conduct" that, in turn, is one of the items in the basic Buddhist "Noble Eight-Fold Path"). It also affirms *ahimsa*, which in fact became more prominent in Hinduism because of Buddhist influence. Veatch cites both a Chinese and a Japanese Buddhist, each of whom applied this principle to medical ethics by exhorting physicians to struggle passionately to preserve life.

In addition, Buddhism emphasizes love, kindness, and benevolence as fundamental to its way of life. This, in turn, is related to its view of the interconnectedness of all things. A human being has two dimensions; one includes the components of body, sensation, perception, will, and consciousness, and the other includes our relationships with nature and other persons. These are also part of us. We cannot find fulfillment if we try to live in and for ourselves as isolated or separated individuals. Thus we are led from a self-centered to a more universal and altruistic perspective.

How does all of this affect a Buddhist ethic concerning transplant operations? One contemporary Buddhist, Tsuji (1988), says that there is no absolute Buddhist position on the matter, but suggests that organ donation for transplantation can express and facilitate the enlightenment process:

An enlightened view of the body and its relation to the whole universe will immeasurably enhance the quality of human life. As medical science progresses, organ transplantation will be perfected and more and more people suffering from various ailments will be helped by enlightened donors. Thus, in the realization of the oneness of humankind and the universe, human beings will share in the suffering as well as the happiness of their fellow beings.

Jewish Perspectives

Abraham Twerski, Michael Gold, and Walter Jacob

An Orthodox Interpretation

Discussion of any issue from a Jewish ethico-religious aspect requires that the author define his orientation. Jewish religion today encompasses a broad spectrum, ranging from the ultra-Orthodox to the ultra-Reform.

Although it may be argued that the Orthodox viewpoint represents only a minority of contemporary Jewry, there is, nevertheless, a valid reason for approaching the issues of organ transplants from this aspect. Orthodoxy is bound by quite rigid guidelines, whereas Conservative and Reform Jewry has greater flexibility, and hence, they may adapt to social or scientific innovations pragmatically. This greater flexibility can obviously circumvent many conflicts.

One caveat is essential: neither this paper nor any other can be used as a guideline for decision-making by a physician, family, health facility, or an ethics committee on any given case involving organ transplant. In Orthodoxy, there is an inviolable principle that any decision involving issues of life and death can be rendered only by a competent rabbinic authority. It is the responsibility of all others involved to supply the rabbinic authority with all the details of a given case to enable him to reach a decision. The decision-making process cannot be delegated.

From: *New Harvest* Ed.: C. D. Keyes
©1991 The Humana Press Inc., Clifton, NJ

It is also important to realize that ethical decisions are as weighty to the Orthodox ethicist as medical decisions are to the physician. For the physician who is confronted with the choice of whether to carry out a potentially hazardous procedure, the issue at stake is the very life of the patient, and the physician realizes that a wrong decision may result in loss of life. Such decisions are, therefore, not made lightly, and made only after all aspects of the case have been considered, usually with consultation from other physicians.

The position of the ethicist is no different. A wrong ethical decision is of grave concern. Ethical/moral decisions may, therefore, require a great deal of study, research, consultation with other authorities, and prayer for divine guidance. It is not possible to feed data into a computer and obtain a decision on a printout, because computers, if for no other reason, cannot pray for divine inspiration. Thus, there is no "cookbook" approach for ethical/moral decisions.

Orthodoxy does not have the capacity of creating ethics or laws *de novo*. Orthodox law is based on Old Testament law, as understood by the Talmud, which is an exegesis analysis and collection of laws compiled from 400 BC to 200 AD. There is no authority in Orthodoxy that can abrogate a law or grant dispensation. Throughout Jewish history, newly arising questions of law and ethics were addressed by scholars who applied Talmudic principles to those situations not directly addressed in the Talmud proper.

Historically, there was a *Sanhedrin* (Supreme Court) in Jerusalem whose decisions were binding on all Jews. With the loss of sovereignty following the Roman conquest of Judea, the *Sanhedrin* was dissolved, and since then, there has not been a single authority whose decisions were universally binding. The Chief Rabbinate in Israel at this time is a governmental rather than an ecclesiastical position, and the opinions or decisions of the Chief Rabbis are no less and no more authoritative than that of any recognized scholar.

The body of laws and ethics that developed in Orthodoxy is referred to as *halacha*. As new situations arise, *halachic* authorities apply accepted *halachic* principles and make inferences from preceding *halacha*, but existing *halachic* principles can generally not be overridden. Not only are Talmudic decisions irrevocable, but also many post-Talmudic authorities have been given the status of infallibility. Their decisions can serve as sources from which one can infer and apply to new circumstances, but cannot be overridden. Thus, although there is some flexibility within *halacha*, it is confined within a rigid framework.

The *halachic* issues that can affect organ transplant are

1. Mutilation of the lifeless body;
2. Deriving any use or benefit from a cadaver;
3. Delaying the interment of any part of a corpse;
4. Hastening the demise of a moribund patient by removal of or tampering with any part of the body; and
5. The obligation of the physician to do everything within his means to preserve human life.

The scriptural antecedents for the *halachic* issues are actually few, and are expanded on in the Talmud. The obligation to preserve human life and hence, for the physician to heal is based on two verses. "Neither shall you stand by idly by the blood of your neighbor" (Leviticus 19:16) and with regard to the restoration of lost property, "you shall return it to him" (Deuteronomy 22:2). The relevance of the latter verse is that if there is an obligation to restore a person's lost property and prevent his suffering a financial loss, there is certainly an obligation to prevent loss of life or health. A third source that is sometimes quoted is "Thou shalt love thy neighbor as thyself" (Leviticus 19:18). Just as one would wish to have one's own life preserved and be relieved of pain and suffering, so is one obligated to these considerations for one's neighbor.

The verse "And surely your own blood of your souls will I require" (Genesis 9:5) is the source for the prohibition of suicide. A derivative of the prohibition against suicide is that just as a person's life is not something he may dispose of, neither is his body. According to this, human life is the possession of God, meaning that a person does not have undisputed title to his own body.

The verse "Surely thou shalt bury him on the same day" (Deuteronomy 21:23) is the origin for requirement of burial, and by application of the principles of derivation, the Talmud also establishes (as Biblical law) that a human body may not be mutilated and that use or benefit of any part of the lifeless body is forbidden.

The scriptural injunction forbidding taking a life is the origin for proscription of euthanasia and for doing anything to the moribund patient that would hasten his demise.

The preservation of human life supersedes all *halachic* prohibitions, except for the three cardinal sins: idolatry, adultery, and murder. Thus, procurement of cadaver organs for life-saving purposes need not pose a major problem, since the various prohibitions cited would be overridden by the supreme requirement to save a life.

Since the kidney is an essential organ for life, the use of cadaver kidneys can be permitted. Some question may arise in regard to corneal transplants since loss of vision is not actually an immediate threat to life. However, the prevailing opinion is that since loss of vision would make one extremely vulnerable to life-threatening accidents, it is considered a threat to life. On this basis, corneal transplantation has generally been permitted.

The greatest difficulty is encountered in heart, lung, and liver transplants because of the need to remove the organ while it is still metabolically active and being profused by oxygenated blood. This means that the organ must be removed from a body whose circulatory system is still functioning. Since a person cannot survive without heart, lungs, or liver, it is nec-

essary that the person be pronounced dead prior to the removal of these organs, otherwise such removal would constitute murder. Since murder is one of the three cardinal sins that
are not overridden by saving the life of another person, there
is no way such organ removal would be permissible until the
donor was pronounced dead. The problem then comes down
to establishing criteria of death, and whether a person can be
considered dead although his body is still metabolically active.

Throughout history, there have been diverse beliefs
among different cultures as to which part of the human
anatomy houses the nucleus of life. The soul has been variously located in the heart, the liver, the diaphragm, and the
pineal body. For *halacha*, the only authoritative opinion is
that derived from the Talmud.

Prior to the advent of organ transplant, there was little
need to establish precise criteria of the moment of death.
This controversial issue is, thus, a rather recent one, and has
been dealt with by various committees and legislative bodies.
The position of Orthodoxy is based on *halachic* decision, and is
not affected by any legislative or judicial position.

The earliest controversy eliciting a definitive *halachic*
opinion as to time of death occurred in 1772 when the Duke
of Mecklenberg ordered that burial be delayed for a minimum period of 72 hours following death to be absolutely
certain that the person had indeed died and was not merely
in a swoon state. As this violated the *halachic* requirement
for prompt burial, Rabbi Moshe Schreiber, known as Hatam
Sofer (1895), an acclaimed *halachic* authority, wrote a
responsum wherein he established three criteria for death:

1. The patient lies immobile as a stone;
2. There is no detectable pulse; and
3. There is no breathing.

When these three clinical signs are present, one can proceed with burial and need not be concerned that the patient
is merely in a swoon state simulating death.

Although Hatam Sofer is a relatively recent *halachic* authority, his responsum has served as a landmark decision for establishing time of death. The obvious difficulty is that this responsum requires absence of pulse, i.e., absence of cardiac activity.

In the Talmud, the only criterion for death is the absence of spontaneous respiration. This may be based on the scriptural reference to human life beginning when, "The Lord God blew into his nostrils a soul of life, and man became a living creature"(Genesis 2:7). Artificial respiration, which permits continued oxygenation of the heart, was unknown until recently. Thus, there is no reference to a hypothetical case in which spontaneous respiration ceases and the heart continues to function.

Concepts such as "brain death" or "irreversible coma" do not have any Talmudic substrate, and, hence, cannot be considered in establishing *halachic* criteria of death.

Currently, there are two schools of thought on criteria of death. The position of the Chief Rabbinate in Israel, which is supported by a number of *halachic* authorities, is that, indeed, the only acceptable criterion of death is absence of spontaneous respiration. However, since simple cessation of breathing has often yielded to successful resuscitation, it must be established that the cessation of spontaneous respiration is absolutely irreversible. This certainty can be achieved, they contend, when there is also absence of cardiac activity, or when the scientific criteria establishing brainstem death have been satisfied. It is, thus, not the brainstem death that is the criterion, but rather that brainstem death establishes the Talmudic criterion of irreversible cessation of spontaneous breathing. According to this opinion, the person is considered dead from a *halachic* standpoint even though the heartbeat is being maintained by oxygenation via a respirator.

A second school of thought, representing many noted *halachic* authorities, takes sharp issue with the Chief Rabbinate position. These authorities contend that presence of cardiac activity, even if maintained only by means of heart muscle oxygenation via a respirator, precludes pronouncing the patient dead. According to this view, there appears to be no acceptable *halachic* way of procuring a heart for transplant given the current state of the art.

An interesting approach has been suggested by an *halachic* scholar, based on the Talmudic ruling that a decapitated person is considered to have died at the very moment of his decapitation, even if there were observed subsequent reflexive muscular twitches. He therefore argues that if perfusion techniques demonstrate brainstem necrosis, this is equivalent to decapitation, and the patient can be pronounced dead. This opinion has been refuted on both medical and *halachic* grounds.

There is ongoing discussion among current *halachic* authorities regarding the criteria of death and hence, heart and liver procurement. As the whole subject is relatively new, there is no universally accepted position yet.

Another issue that arises in the case of kidney transplant is that of procurement from a living donor. Given the *halachic* position that a person does not have undisputed title to his own body, is it indeed permissible for a person to donate an organ while he is still alive? Is it his to donate? If donating the kidney presents any risk to the donor's life, is he permitted to expose himself to this danger? Would donating a kidney constitute a transgression of the biblical injunction against self-harm, "Take heed to thyself and keep thy soul diligently"? (Deuteronomy 4:9)

This issue appears to have been resolved on the basis of a passage from the Jerusalem Talmud, which concludes that a person is obligated to place oneself even into a possibly dangerous situation to save another's life. The majority of *halachic* authorities, therefore, concur that a small risk may be undertaken by the donor if the chances for success to the recipient are substantial, and that even a major risk may be undertaken if the recipient would die without the transplant.

The permissibility of transplanting fetal tissue, as has been done with adrenal tissue for relief of Parkinsonism, is controversial in *halacha*, with opposing views expressed by two nineteenth century authorities.

Recent advances in infertility treatment have been addressed by *halachic* authorities. Artificial insemination using donor semen is generally prohibited, but artificial insemination using the husband's semen is permissible. In vitro fertilization appears to be *halachically* unobjectionable, although some authorities believe it to be morally repugnant.

Surrogate motherhood, wherein the sperm and the ovum of a married couple and the fertilized zygote is implanted in the uterus of a host female, is currently under analysis by *halachic* authorities, and at the time of this writing, nothing definitive has been published.

Implantation of a mechanical heart or that of an animal is dependent entirely on the perfection of the procedure. If the procedure holds sufficient promise of success, it falls within the guidelines concerning all other hazardous procedures.

Other Jewish Interpretations

A Conservative Response

A Conservative rabbi would agree with most of the preceding essay. Only competent Rabbinic authorities can decide the permissibility of medical procedures such as organ transplants, fetal tissue research, and the new reproductive techniques. They can do so only after a careful study of the Jewish legal precedents known as *halacha*.

However, a Conservative rabbi has a different understanding of *halacha* than his Orthodox colleague. For the Orthodox, the *halacha* is God's word, and as such, is inviolable. To Conservative Jews, *halacha* is a human interpretation of God's word. Thus, any particular decision must be studied within the context of the time it was promulgated.

Conservative authorities would agree with the preceding essay in this chapter that a human body is precious and the mutilation of the dead, delay in burial, or deriving benefits from the dead are forbidden. However, all these prohibitions may be set aside to save a life. Therefore, these are not a consideration in organ transplants. In fact, all the laws in Judaism (except three: prohibitions of murder, incest, and idolatry) can be set aside to save a life.

The problem with organ transplants is the determination of the exact moment of the donor's death. Contemporary medicine has changed the criteria for establishing death. To quote the late Rabbi Seymour Siegel (1976), the leading Conservative thinker on bioethics, regarding these new criteria of death:

> The basic Talmudic definition is that death occurs when respiration ceases. The cessation of heartbeat was also taken into consideration. There are many today who believe that a better definition of death should be based on the cessation of brain activity as evidenced by the absence of EEG's (brain waves). When the brain waves stop, spontaneous breathing and heartbeat are impossible, and therefore the patient can be said to be dead even though his systems may be moving by means of machines. It was felt that when the rabbis defined death by referring to the circulation and breathing, they were reflecting the best scientific information available in their day, but now that we have means to measure the activity of the brain, the organ which is the central mechanism for the support of life, a new criteria for death should be adopted.

A number of secondary issues were raised in the preceding essay. Conservative authorities would agree that a person may place himself or herself in a possibly dangerous situation to save a life. Thus, kidney transplants from a living donor are permissible. The new reproductive techniques such as artificial insemination with donor sperm, in vitro fertilization, surrogate motherhood, and embryo transplants raise numerous *halachic* issues. These were studied at length in Gold's recent book (1988), where it is indicated that there is room for permissive rulings based on Talmudic sources, particularly in light of the certainty of procreation as a commandment in Judaism.

Life is a gift from God and is, therefore, of ultimate value. When a medical procedure can save or enhance a life, the rabbis should search for a permissive ruling.

Reform Response

Reform Judaism makes its ethical decisions within the framework of the Jewish Biblical and rabbinic traditions. Decisions in each age also combined tradition with the secular knowledge available in that period and the condition of Jewish life.

The scholars of the Mishnah and Talmud were aware of medical science and research done in their day or earlier by Greek physicians. They used that knowledge as a basis for many of their decisions. Subsequently, many of the greatest rabbinic scholars and Jewish essayists were also physicians. The best known among these individuals is Moses Maimonides (1135-1204). He wrote extensively on medicine and made use of this knowledge, particularly in his famous code, *Mishnah Torah*. This involvement of Jewish essayists in medicine and ethics has continued to the present day.

The Reform point of view seeks to continue the development of Judaism and to combine the insights of tradition with the latest knowledge gained by medical research and practice. We are cautious in experimental areas, but are willing to use efforts that will help patients and especially those that will save their lives. In Solomon B. Freehof's 1968 decision, a positive step toward the justification of transplants was taken, with the caveat that nothing may be done to hasten the death of the dying individual. A later decision by myself accepted the criteria of death set by the Ad Hoc

Committee of the Harvard Medical School in 1968. These criteria would make it possible to transplant a heart, a lung, or other vital organs while oxygenated blood is still circulating once brain death has definitely occurred.

Kidney and corneal transplants present us with few ethical problems. Kidneys from living donors still enable the donors to function perfectly well. Corneas can be taken only from those who have died. The sale of kidneys is, of course, prohibited by our tradition.

Artificial insemination is generally permitted, using either the husband's semen or semen from another source. The Orthodox prohibitions against a source unknown to the mother are primarily due to a possibility of incest, but we view that possibility as extremely remote, and therefore, have no problems with it.

Surrogate motherhood raises ethical questions about the economic nature of the transaction and the maternal feelings that may develop despite all wishes to avoid them. We would, therefore, hesitantly permit this, but continue to study the matter.

As we consider such issues, we should keep in mind the dictum, "God created medicines out of the earth and let not discerning man reject them" (Ben Sirah 38:4), and the injunction of Maimonides that "the well being of the soul can be obtained only after that of the body has been secured."

Christian Perspectives

David Kelly and Walter E. Wiest

Roman Catholic Interpretation

Christianity, like Judaism, proposes a people of God, not just a loose collection of private individuals. The human person is created in God's image, chosen, and ordered to life with God through the grace of Jesus Christ given in the Holy Spirit. This act of creation, choosing, and ordering is not primarily an individual event. Humans are social beings. This does not mean, however, that the rights of the individual can be neglected, but rather that each person finds value in the interconnectedness and interdependence of human society. In Roman Catholic moral theology, the fundamental themes of human dignity (respect for life) set the context for moral decision-making in the area of health care. The interpretation of ethical issues in transplantation that follows is a continuation of the historical development and foundational themes of Catholic medical ethics discussed by Kelly (1979,1985).

Organ Transplantation in Catholic Medical Ethics

Ethical concern about organ transplantation arose largely in the 1940s and 1950s. At that time, the main issue was the problem of mutilation. Could a person "mutilate" him or herself for any purpose? The principle of totality was generally applied here. According to this principle, a part of the human body could be sacrificed for the good of

From: *New Harvest* Ed.: C. D. Keyes
©1991 The Humana Press Inc., Clifton, NJ

the whole body. By the mid-twentieth century, the principle of totality had become an integral part of Catholic medical ethics. This principle has been criticized in recent years, since it was originally limited to the physical good of the individual physical body. Mutilations were permitted only if an organ was diseased so that the physical good of the same physical body required the operation. Thus, a gangrenous leg might rightly be amputated or a cancerous organ removed. Blood transfusions and skin grafts were usually considered nonmutilating, and hence, were permitted. But on this principle, transplantations of cornea or kidney were likely to be forbidden, and many Catholic moral theologians did indeed forbid all "mutilating" transplants.

A turning point came in 1944 with the publication of Bert Cunningham's dissertation, *The Morality of Organic Transplantation*. His central thesis was that organ transplantation is indeed licit according to the principle of totality, if this principle is taken to include not just the individual donor's body, but also a social reality, the Mystical Body of Christ. By this term, Cunningham means more than the Catholic Church. He means the unity of humankind as a race created by God and redeemed as a Body by Christ. Given this unity, a person may licitly mutilate him or herself for the good of the neighbor.

Cunningham gives examples of procedures his thesis would permit and some of them are extreme. For example, he allows for the transplantation of a cornea from a living donor, resulting in considerable loss of vision, and even the corneal transplant of one eye from a one-eyed donor resulting in total blindness. In this last case, the donor is presented as a convict sentenced to life in prison, and thus, "not needed by anyone." There is a lack of respect here for the individual person that arises from the thesis that individuals are ordinated to society as parts to the whole. But Cunningham's principle of totality enabled him to allow for the possibility that organ transplantations might be morally right. Cunningham's thesis was hotly debated by Catholic moralists during the forties and fifties. Many European moralists continued to oppose all organic transplantation. Most American moralists and some Europeans accepted Cunningham's conclusion that organ transplantation was licit, though many of them expressed hesitation at the lack of caution or proportion clear in some of Cunningham's judgments. They tended to allow operations that do little if any harm to the donor and are of great benefit to the recipient. Since the Church Magisterium (the Pope and the Bishops) had not made any pronouncements directly forbidding organ transplantation, these Catholic moralists were willing to permit the procedure. They disagreed about whether or not the principle of totality could be applied. Some argued that the principle must continue to apply only to the individual body; these justified organ transplantations on the basis of charity. Others argued that, with proper safeguards, the principle of totality might be extended in a way similar to Cunningham's proposal.

By about 1960, the debate had died down. Most Catholic moralists by then accepted the moral rightness of at least some organ transplants, and in the intervening years Catholics have joined others in seeing this procedure not only as a morally correct act, but indeed as a laudable one to be supported by public policy. There is still hesitation, however, among Catholic teachers about the difficult question of the relationship of the individual (live) donor and the recipient. Catholic teaching rejects the idea that the human individual is of no worth apart from what she or he can contribute to others. Organ transplantation is morally right because it contributes to and is in keeping with the created and saved dignity of the donor as well as that of the recipient.

McCormick (1975) proposes a different understanding of "totality" as applied to organ transplantation. This Catholic theologian extends the notion not so much toward society as toward the total personal good of the donor. We are spiritual as well as physical beings. Donating an organ, although it involves physical injury, might well be counted a spiritual good, a gift of love to another. The spiritual good may outweigh the physical harm. However, this does not mean that any and all organ donations are morally right. In some cases, the loss to the donor might outweigh any benefit to the recipient. There must be a correct proportionality of benefits and harms if donation is to be morally right.

Though no Catholic moral theologian today argues for a general rejection of organ transplantation, there are still a number of very vexing questions that call for morally adequate answers. For instance, there are questions about the cost and distribution of scarce resources. It is clear that society should use its resources justly and ought not spend so much on procedures benefiting only a few that it neglects to do what it should to benefit many. We are called to participate with one another in working for the common good. It would be arrogant, however, for moralists to claim that easy specific conclusions can be drawn from these obvious premises. We must hesitate before rejecting expensive procedures simply on the basis of cost. It may be theoretically true that the money spent on transplant surgery could help large numbers of the poor or the starving, but the complexities of today's global economy make such judgments questionable at the practical level. How can we know that it is expensive organ transplantations and not educational costs, military costs, entertainment costs, church costs, or other allocations of society's resources that deprive the poor of needed care and support? Yet, the complexity of the question cannot cause us to give in to the temptation to ignore it.

Though there is no clear Catholic teaching on specific policies in this area, there is a growing emphasis on the need to work for the wider good of society and to show a preference for the needs of the poor when making personal and social decisions. At the very least those medical procedures that are of doubtful benefit to the individuals who receive them or to the relatively small population of potential recipients are open to moral criticism if it can be reasonably foreseen that the time, effort, and cost spent on them could be put to better use elsewhere. Perhaps this combination of factors applies to artificial heart implants, at least for the present. As argued elsewhere by Kelly (1987), this procedure ought to be limited to those experimental attempts necessary for developing a better, longer lasting, and more cost effective device.

Organ Procurement and Determination of Death

The question of when death occurs is not in itself an issue of organ transplantation. Indeed, it is morally and medically better to distinguish, even to separate, the two questions as much as possible. Yet it is a fact that one of the problems connected with cadaver organ procurement is determining that the donor has died.

Organ procurement involves a number of policy questions apart from the determination of death. Chief among these is who controls the body of the deceased and who decides whether or not organs can be taken. There are a number of different policy options, and governments have adopted one or another of them. Some argue that cadaver organs ought to be considered the property of the state, or at least that the common good should require by law that they be available for transplantation, regardless of the wishes of the person before death or of the relatives after death. Opposing this is the position that organs be taken only if relatives spontaneously volunteer them or if the now dead person has made his or her wishes to donate clear during life. Between these two positions is the policy generally known as required request, now adopted by a number of states in the US. This policy requires by law that the relatives of a dying or newly dead person be asked about possible organ donation, assuming, of course, that transplantation is medically possible. If the deceased carried an organ donor card, or otherwise made his or her wish to donate known, this wish is seen as valid consent.

There is not any clear position among Catholic pastors and theologians on this issue. Scholars are apt to emphasize the importance of giving to others and the freedom of choice that this necessarily entails. On the other hand, Catholic social ethics recognizes that there are times when individual choice must cede to the common good. On this issue, Catholic scholarship is likely to support a moderate policy between state control and purely spontaneous donation. There is no reason why required request laws should be opposed by Catholic moral teaching as they include both the important freedom of choice necessary for giving and the valid needs of the common good for an increase in available organs.

One question often raised here concerns the Resurrection of the Body. Traditionally, the Catholic Church has supported burial rather than cremation to express the belief in bodily resurrection and to respect the created and saved goodness of the body. Yet it is has become clear lately that cremation is neither a sign of lack of respect nor a way of denying faith in the final resurrection. Today, Catholic pastors and theologians no longer oppose cremation. They point out that the Resurrection of the Body cannot depend on the former body being intact; no one's body remains intact for long after death.

Similarly, there is no reason why organ donation after death should be seen as denying a hope for resurrection. Indeed, organ transplantation should be seen as quite the opposite. It is a symbol of human and of Christian communion, with Eucharistic connotations. Here is a body given for others. Though it is not a difficult or heroic gift—once I am dead I do not need "my" organs any longer—it does allow us to share God's gifts by letting what was a part of us give life to others. It is in keeping with this deeper theological understanding that Catholic teachers no longer oppose organ donation by living donors, as we have already seen. The same understanding supports cadaver donation, and Catholics have joined others in educating all of us about the moral and religious beauty that can be present in organ transplantation.

Since "fresh" organs are better than "stale" ones, a quick means of diagnosing death is desirable. The best organs are often those of young persons who have died with head injuries, resulting in death to the brain but with little direct injury to the rest of the body. In the US thus far, brain death criteria require the death of the entire brain, including the brainstem. No organs can be taken from a patient until the heartbeat and respiration have ceased, or, when heartbeat and respiration are sustained by machines, until total brain death is determined.

The controversy today continues about whether we should not instead turn to "higher-brain" death criteria and recognize that those whose neocortical functions have completely and irreversibly stopped are indeed humanly dead. Could we not take their organs? Though there is no clear Catholic position on this issue, either side could be supported. Janssens is aware of both sides of the issue. This Catholic theologian (1983) argues that "the death of the cerebral cortex means the end of the historical presence of the human person...[it] is the elimination of the biological conditions of all human potentiality; it marks the historical end of man as man." Janssens also claims that, whereas scientific data support the contention that cortical death is the real death of the human person, sociocultural contexts argue otherwise, at least for the present.

On the one hand, it is clear that the human person is more than mere animal, and Catholic ethics has for a long time recognized that there is no obligation to sustain life when the benefits of doing so are outweighed by the burdens of the medical treatment. On the other hand, Catholic theology would be slow to say that the entire human person is merely contained in the functioning of the brain. In addition, there are dangers involved in changing the social understanding of death. How would society cope with burying bodies still breathing, or with acting to stop respiration in what had been a human person? It is quite possible that Catholic opinion will change toward permitting neocortical determination of death. Clearly, persons in the persistent vegetative state who breathe on their own but who do not and will never have any awareness of self or other, conscious or subconscious, need not be kept alive by medical means. They may and ought to be allowed to die. But so far, most Catholic moralists are hesitant to say that they are already dead. Still, Catholic theology can contribute to the consideration of such questions even though it has as yet no definitive answers.

The question of anencephalic newborns as potential organ donors is a particular instance of the question of determining death. Anencephalics are born without any cortex but many are born with the brainstem intact. They are "brain-absent." They all die very young, most within days, almost all within weeks of birth. Because the brainstem is intact, they may well breathe on their own during their short lives. At present, physicians must wait until their breathing stops before taking organs, since the law requires total brain death, including the death of the brainstem. This is a medical problem, since their small organs deteriorate very rapidly without oxygen. Some suggest a change in the law adding anencephalics to the category of those who are indeed dead. Thus far, the Catholic consensus would seem to be that this should not be done. Perhaps erroneously, official Church opposition to abortion often operates in this context, as some fear that a change in the law would open the doors to further defense of permissive abortion laws. Others might argue that the anencephalic is, indeed, a very special case; no human life is actually or even potentially present here. It is possible that Catholic opinion on this might change in the future. It seems that laws adding anencephalics to the category of those declared dead could be specific enough to prevent abusive extension to other categories. Still, we ought to be very hesitant here and aware of social dangers.

There is one final aspect of this problem that ought to be addressed. It must be emphasized that the issue of determining death is not as such an issue of organ transplantation, even though that has often been its context. We ought not change our laws about determining when a person is dead merely in order to facilitate organ procurement. Thus far, American law and policy has not really changed what it means to be dead; it has merely changed the criteria for determining that death has occurred. To add anencephalics or other neocortically-dead bodies to the list of the legally and medically dead would change what it means to be dead. There could be a backlash against it that would work to the detriment of organ transplantation. This could change in the future, of course, as people become more open to the notion of neocortical death criteria. But at least for now, caution should rule.

Brain Tissue Transplants

The question of brain tissue transplants involves special questions in two areas. First, to what degree might the transplantation of brain tissue affect the self-identity of the recipient? Second, what about transplantation of fetal brain tissue for treatment of ailments like Parkinson's disease?

Transplantation into the brain to treat Parkinson's disease and the speculative types of transplantation to restore functions mentioned in an earlier chapter of this book do not attempt to transfer personality from donor to recipient. Furthermore, Keyes claims there that the present limits of our knowledge of the brain provide no grounds for speculating that tissue could be effectively transplanted from one adult brain to another. Transplantation of an entire

brain from donor "A" to recipient "B" is even less thinkable
for now at least, and borders on the outrageous or even the
ghoulish. This may not always be true, of course, as anyone
aware of what medicine has been able to achieve well knows.
Catholic theology argues that the human person is enspirited
flesh and embodied spirit. The spirit does not reside in the
brain; yet the brain is clearly the one organ without which
all human function is impossible. This does not mean, however, that a totally disembodied brain is necessarily a human being. If such a thing could ever be made to exist,
perhaps even by massive transplantation and implantation
of organs and artificial devices, I would be slow to call it
human, though it would arguably be rational, even spiritual. Such speculation helps to prepare us for what might
arrive in the future even if it is of little practical importance
for the present.

Of more practical import is the question of fetal organ
and tissue transplant, including the transplantation of brain
tissue for treatment of Parkinson's disease. There is some
controversy over this in Catholic ethics. The official Catholic position on abortion is well known. Catholics would
rightly oppose any attempt at "growing" fetuses simply to
use their organs after abortion. This has been done and is
being proposed but it would reduce the potential human
person to a mere thing. Even for those Catholics who do not
agree with the official church teaching banning all direct
abortion on the basis that the fetus, even very early in
gestation, is a fully human personal life, the deliberate
creation and killing of the fetus for transplant purposes
would be seen as heinous.

Less clear, however, is organ transplantation from a
fetus that has been aborted or has spontaneously miscar-
ried. On one side are those who argue that permitting
transplants will increase the likelihood of abortion, and that
this is abuse of the fetus or the newly aborted child. On the
other side are those suggesting that once the abortion has

taken place and the aborted fetus is dead, the organs and tissue may be taken as with any other dead person. This side would argue against any techniques used to keep the organs usable if these techniques would further endanger the life of the mother or of the fetus, and would insist of course that any child born alive be treated like any other person.

Sexually Related Transplants

Concerning the procedures of artificial insemination, in vitro fertilization, surrogacy, and related questions, Catholic pastors and theologians have engaged in a long investigation and debate. Though the issues are quite complex, and though the conclusions Catholic scholars and leaders have reached concerning them involve methodological questions of long history and much controversy, it is possible in this brief essay to summarize the main points. The official Catholic position on these issues insists that (1) each and every sexual act must in itself be open to the possibility of procreation (the basis of the official rejection of artificial contraception), and (2) no procreation may ever be done in the absence of a sexual act (artificial insemination, therefore, is forbidden, except in the context of a conjugal act such as sperm, taken after sexual intercourse, being inserted into the cervix as a possible aid to fertilization).

With these principles established, the official teaching of the Church, as expressed, for example, in the recent Vatican document on Reproductive Techniques (1987), goes on to forbid even more totally such techniques as artificial insemination by third party sperm donor, which is said to be an injustice to the exclusivity required in marriage, and "surrogate motherhood," which adds to the "technical adultery" a host of possible psychological, legal, and social problems.

It should come as no surprise to the educated reader that this official position has received much criticism from within the Catholic community. This has been most recognized in the area of birth control, with which many (per-

haps most) Catholic moral theologians and a large number
of priests, along with the great majority of lay people, have
publicly or privately disagreed. They argue that requiring
each act to be open to procreation is to place too much
emphasis on the act itself and not enough on the personal
meaning of married sexuality.

Similarly, in the issues that concern us here, a good
number of moral theologians have argued that, in at least
some cases, artificial insemination is morally permissible.
This is clearest in the so-called "simple case." The "simple
case" is when a married couple, using the sperm of the
husband obtained by masturbation or other means, attempt
to conceive a child by artificial insemination with or without in vitro fertilization. It is hard to see how the official
position, which rejects this, can be supported on theological
grounds. There is no adultery here, not even a technical
kind. The procreative process is kept humanly within the
marriage. Medical technique is used to replace the normal
process only because the married couple is incapable of
conceiving without this help.

As further complications, such as a third party sperm
donor or a "surrogate" mother, are added, however, even
those Catholic thinkers who disagree on the "simple case"
are apt to agree with the Church's official position, though
some hesitation remains. There are indeed serious psychological and social dangers in surrogacy and in AID, dangers
for the child, for the spouses, and for society as a whole.
Though some women might possibly find the gift of surrogacy a fulfilling and meaningful endeavor, it is at least
equally likely that this would become a further means of
oppressing women, especially poor women. The spouses
need psychological strength to go through the processes
involved, and children might suffer the ill effects if parents
or others are insensitive to their needs.

All of this becomes even clearer when the issue is large
scale genetic experimentation and engineering. Here, the

Catholic position is less debated and more a consensus. Genetic engineering, which is a kind of surgery aimed at treating a disease in an individual person, is seen to pose no particular problem as long as all the usual safeguards required of any medical treatment are provided. These safeguards include the insistence that new techniques be approached with caution until it is clear that the likely benefits outweigh the burdens and the side effects. Informed consent must not be overridden by a desire for a scientific breakthrough and personal prestige. Problems specific to genetic engineering arise when techniques are suggested that would change the gamete cells so that future people would be "different, engineered" with special qualities and characteristics. Cloning, which is the futuristic technique whereby an adult person's genetic code would be reproduced exactly in one or more babies, is fraught with technical and moral problems making its rejection imperative. Fortunately, it is not yet possible, and may never be. To go into these areas in detail would require that this essay begin anew, rather than come now to a conclusion.

Roman Catholicism at its best offers theological and ritual symbols that can serve as a matrix for approaching the difficult issues with which this essay, and this book, are concerned. Though some of the questions are of great complexity and though answers may not be easy, those of us who base our search on religious and liturgical symbols of faith are thereby aided in our work of collaboration with all women and men who are aware of, and working toward, morally right decisions in health care ethics.

Other Christian Interpretations:
A Protestant Response

Although no one person can speak for all Protestants, they are united these days by currents of thought that cross denominational lines. This essay reflects beliefs that are shared by many American Protestants today concerning certain specific ethical questions raised by transplant operations.

It is a happy fact that these days, many Protestants and Roman Catholics find themselves in much larger agreement than at any time since the Reformation. This is especially true in regard to social issues. Of course, such issues create their own divisions, but it is significant that, often, both Protestants and Catholics can be found on both sides. Traditional differences remain, but there is a bridge of unity. We must add that, although we cannot presume to speak for other branches of Christianity such as Eastern Orthodoxy and some types of Anglicanism, we hope the beliefs we share with them provide enough common ground for agreement on many of the issues treated here.

Roman Catholic ethicists have taken the lead among Christians in addressing the ethical questions raised by modern medicine. Protestants entered the lists mainly within the last thirty years. By that time, Catholic thinkers had already contributed substantially to current thought. Even secular medical ethics had absorbed Catholic concepts such as "extraordinary means" and "double effect." The Catholic voice had also been heard in defining human life, when it begins (contraception, abortion) and when it ends (or should be allowed to end), and on many other issues.

Perhaps the first significant Protestant contribution came from Joseph Fletcher (1954) in his book *Morals and Medicine*. Fletcher maintained that it is not good enough to judge what is ethically right or wrong simply by applying certain moral principles and rules to particular cases. Rules have their place, but cases or situations vary widely. What we need, said Fletcher (1966), is "situation" ethics. The basic ethical question is this: In these particular circumstances, what can be done to achieve the best outcome, the greatest good?" Any rule might have to be bent or broken in some cases. Any act could be right in some situations. The end justifies the means. Fletcher's way of thinking was later challenged by Paul Ramsey (1967,1970). Ramsey reacted with a much stronger emphasis on principles and rules. The end or result cannot justify all means. Some things should not be done regardless of the consequences. In this regard, Ramsey's thought has come much closer to that of Roman Catholic ethicists; *see* Curran (1973) for a comparison of his thought with Ramsey's.

Issues like these concern us all, even those who are purely secular in their views. For instance, should we subject sick children to medical experiments that cannot promise them a cure (and might even harm them) on the grounds that this contributes to the search for a remedy that could save millions of other children? If we follow a "situation" ethic, we might well answer, "Yes." The experiment seeks "the greatest good of the greatest number." The end justifies the means. If we follow a "rules" ethic, we would likely say that such experiments are always wrong. The value of a human life has priority, especially an innocent and helpless one. We are obligated to protect it regardless of other considerations. An adult might volunteer for such an experiment; children are not mature enough to volunteer and no adult has the right to volunteer for them.

Transplant operations raise questions that are often answered with a bias toward one of these two ethical points of view. For instance, should an anencephalic infant be allowed to die and its organs and tissues be used for transplant operations? One who emphasizes rules would be concerned first about the anencephalic infant itself. Is it a human being? Has it rights that should be respected? When these questions are asked first, one's inclination is to say "yes" or at least to lean toward this answer just in case it should be affirmative. If so, then the infant cannot be so used. One who is a "situationist" would be more inclined to approve the use of anencephalic as donor. Its organ might well enable an otherwise normal infant to have a reasonably full and satisfying life. The anencephalic itself has no hope of life beyond infancy. In any case, the greater good should prevail; the end justifies the means.

Both Fletcher and Ramsey reflect important elements of Protestant ethics. Principles are necessary to morality, but they sometimes mislead or conflict. For example, medicine is meant not only to protect or preserve human life, it is also meant to relieve or prevent suffering. But in the last stages of a fatal disease, medications used to relieve pain sometimes also shorten life. Is this murder, or compassion?

At times, the problem is that we can hold too rigidly to a relatively minor rule or law, and thus, violate a larger and more comprehensive moral principle. This was Jesus' concern in his reaction to certain rules about Sabbath observance in his time. Should the rule against doing any work on the Sabbath be taken to mean that diseases should not be treated on that day? Is medical "work" prohibited? Jesus' response was that Sabbath rules were made for the benefit of human beings ("The sabbath was made for man, not man for the sabbath." Mark 2:27) and that a too legalistic regard for particular rules could defeat the very purpose of the rules themselves.

Fletcher has a point. Rigid adherence to rules can lead to inhumane judgments and actions in particular cases. Ramsey also has a point. Ethics without rules can encourage us to treat some moral obligations too lightly. Even when we rationalize, we pay our respects to rules of behavior. Justifications usually appeal to rules that define some sorts of actions as right and some as wrong. Protestant ethics do offer some principles that suggest guidelines for medical practice.

Basic Principles

Martin Luther insisted that we are saved by the grace (love and forgiveness) of God and not by living up to the requirements of moral laws. However, he was horrified by the "anti-nomians" who took this to mean that Christians need obey no laws at all. His point was mainly about motives; if our hearts are right, we will obey the laws of God because we want to, not because we think we can earn God's approval or avoid Divine judgement.

Protestants agree on the principle that every human life has intrinsic value, which is bestowed by God. But love of God and love of neighbor are joined and we are all neighbors. As Calvin (1559) stated, "since God has connected mankind together in a kind of unity, every man should consider himself as charged with the safety of all."

This means that we affirm the importance of the individual while believing that individuals are not meant to live just for themselves. Humans are also intrinsically social. We have sympathy for Richard McCormick's argument that the "total personal good of the donor" can be enhanced by giving to others even if the gift costs something. We fulfill ourselves in giving to others.

It is important to say, however, that damage done to one's body is an important consideration. There have been "dualistic" religions and philosophies that emphasize the soul in a way that disparages the body. This attitude has affected Christian thought in the past. In this century especially, biblical scholars have shown that such dualism is foreign to biblical faith. We are "psychosomatic" creatures; a unity of body and spirit. Our bodies are part of what and who we are. That is why donors sometimes feel they have somehow become part of another person and recipients can feel invaded by a foreign element. We must be careful what we do to bodies.

On the other hand, our bodies die. We are not intrinsically immortal. Protestants believe (at least traditionally) in God's gift of eternal life, but eternal life is not that of an inherently immortal soul released from a mortal body. It is a transformation of our whole being, body and soul. As far as this life is concerned, we are mortal in both body and spirit. Therefore, we all face a time when we must accept death. Further, we all should be allowed to die with dignity. Respect for individuals includes respect for their dignity and freedom.

This is part of the love we are to extend to our neighbors. We have an obligation not to force on patients some added months (or even years) of "life" that is robbed of elements that make life really human and that can only delay the inevitable end.

Such thoughts are not meant to oppose or denigrate what modern medicine can do to save and extend lives. We join with Catholics in believing that humans are created in the "image of God." Part of that image is that we are "cocreators" with God. We cannot create a world out of nothing but we can create the "secondary environment" of human culture and civilization, the totality of human arts and sciences. Medicine is part of all of this. Whatever the beliefs of its practitioners, we accept it as a gift of God.

There is one reservation: with our great gifts, we can be seduced by arrogance. We can come to think that we should be allowed to do whatever we can do. As Jacques Ellul (1954) has observed, the tendency of technology is to assume that whatever can be done will be done. It is easy to see the immorality of uses of medicine such as those employed by Nazi doctors in concentration camps. It is not so easy to see the immorality of medical professionals who want to try every new technique they are sure will work or who want to keep patients "alive" by any minimal definition in spite of the damage done to dignity or the cost in suffering. There are moral limits to our "co-creative" powers. With all this said, we now turn to further comments on particular issues raised by transplant operation.

Procurement

Procurement from Living Donors

Some organs and tissues can be procured from living patients. The damage to the donors varies. A person with one kidney can live a long life. A person who donates an essential part of her/his eye will live half-blind. Obviously, we cannot demand such sacrifices. We can encourage those that involve relatively less loss to the donor. Individuals might want to sacrifice their very lives for another person, but this cannot be part of ordinary medical ethics. Doctors must protect the lives of donors as well as recipients. On the other hand, it may not be ethical to reject the offer of a mother to give up part of her sight in order to grant sight to her son or daughter. Surgeons who have compunctions about this should not be forced to perform the operation, but should they be allowed to block it? If other surgeons can be found who are willing, let one of them proceed.

Procurement from Dead Bodies

Some organs (hearts, obviously) can be procured only from the bodies of the dead. Modern medicine, though, has devised ways of keeping bodies going by mechanical means. When is a person dead? When can a patient be considered eligible to be donor of a heart?

It is generally agreed that we should not define death just to meet our need for transplantable organs. As the preceding Roman Catholic interpretation says, the two issues are separable. Yet, what we decide about the one will affect the other. The current definition of "brain death" and the criteria for determining it are now accepted by law in most of our states. This is good. It relieves us of the need to keep patients on mechanical supports long after they could have stayed alive otherwise. When their brainstems can no longer keep their hearts and lungs operative, they are dead. The machines can be disconnected. They are potential donors of essential organs. This is quite in keeping with a "psychosomatic" view of human beings. Soul or spirit is linked with body.

Of course permission must be given. Even after death, we associate bodies with persons. They are to be treated with respect and cannot simply be commandeered. Since there can be problems about permission, it is well for us to make a statement in writing, in advance, if we are willing to have our organs and tissues used for transplantation after we have died. Failing that, there are other possible procedures. These have been presented in Section 1 of this chapter, and we agree that "required request" laws seem the fairest solution. Doctors can be required to ask the family of a patient who has died whether the patient's body can be used for donation.

But what should we do about patients who are irreversibly comatose? They are not legally brain dead, yet they may be prime candidates for donation of organs and tissues. Might the definition of brain death be extended to include them? Protestants might approve of such a move (though not all will) on the basis of our psychosomatic view of human beings. The spirit is as essential as the body. A body without consciousness is not a human being. Therefore, where the basis for consciousness is lacking, bodily functions need not be supported indefinitely (including artificial feeding and hydration, though there are further questions about this). There is no human life at stake.

This applies even more to anencephalics. Whereas a comatose adult has lived long enough to acquire living relationships, to love and be loved, an anencephalic infant or fetus can be loved but cannot respond, and cannot live very long.

Although it may seem brutal to judge such individuals dead (i.e., not really human), such judgment seems better than two alternatives: (1) If the comatose are judged alive and fully human, then they must be kept going indefinitely (in Karen Quinlan's case, for ten years) without any hope of conscious existence. This is a travesty, and, of course, a costly one. (2) To judge such individuals dead (i.e., not humanly living) is better than to say we can use them as donors because they lack an acceptable "quality of life." "Quality" admits of degrees. We cannot justify using mentally retarded persons as "transplant fodder" any more than we can justify (as the Nazis did) any medical manipulation of humans whom some distorted ideology might view as inferior and subhuman. Instead, we take the absence of a functional neocortex as decisive. Without it, there is no human being.

Allocation

The supply of transplantable organs is limited. We then face two questions: (1) Who gets the transplantable organ? (2) How much of our health care resources should be devoted to these expensive operations, as compared to (a) other medical needs and (b) other social needs? These questions are addressed in other essays included in this volume. Whereas Protestants and Catholics have concepts of justice that apply to such questions, neither of us have as yet achieved a consensus. As suggested in the preceding essay, it is difficult to decide what is ethically right in this area. We shall all be thinking further about the issues.

Brain Tissue Transplants

Since the brain is the basis of consciousness, what effects might transplantation of brain tissue have on our humanity, our self-identity? So far, transplantation into the brain has involved only cells and tissues and has not appeared to affect self-identity, although further research should be done in this area. The pressing ethical issue at the moment has to do with the use of fetal tissues as therapy for Parkinson's disease. Although different views of abortion and of the status of fetal life are involved, most Protestants would likely agree with the preceding Roman Catholic interpretation that (1) once a fetus has died, through either abortion or miscarriage, its tissues may be used, with permission, and (2) that no fetus should be "grown" simply to be aborted and used for this purpose.

As for whole-brain transplants, the question is not so much whether they can be done as it is whether we should try to do them. Although brains have not been transplanted, the heads of animals have been transplanted to other animal bodies, with some success. Can we really assume that the experimenters have no notions about how the techniques might be applied to human beings? Once a thing becomes technically possible, it is harder to resist. It is not too soon to consider the ethical questions.

No one knows what sort of person would be created by transplanting one brain (or head) to another body. We are very complex selves. While our brains are essential, so are our glands, our faces, our stature, our physical coordination. We "are" our relationships, also, with family, friends, colleagues. A transplanted brain would carry with it some characteristics of one person but would have to be merged with characteristics of another person. The result could be disastrous, an unresolvable crisis of self-identity, a troubled and tortured creature. The debate about this is barely beginning but we must at least ask whether the motive behind any such efforts is really that of saving lives or whether it is rather the desire to do whatever can be done by science and technology.

Sexually Related Transplants

Since most Protestants accept contraception, we do not have to face the questions raised by Catholic teaching on this point. On other issues, there is much debate and difference of opinion. At least this much can be said, all too briefly:

1. Artificial insemination by sperm from the husband (AIH) should be acceptable. Protestants traditionally have placed more emphasis than Catholics on the sexual act as an expression of the love between husband and wife, apart from procreation. This makes us freer to accept artificial conception for couples who cannot conceive naturally, despite the fact that the process separates procreation from the physical "act of love." In fact, we can understand AIH as itself a loving act and a fulfillment of marriage.

2. Insemination by sperm from a third party (AID) is questioned by some, but especially where the third party remains anonymous, there is no adultery involved and no relationship that intrudes on the marriage. In effect, the husband "adopts" the child.
3. Surrogate motherhood is a vexed issue on which there is probably as much divided opinion among Protestants as within the general public. We can at least agree with the summary in the preceding essay of the most important ethical questions that must be considered.
4. On genetic engineering, we can agree that (a) efforts to correct genetic defects (sickle cell anemia, Tay Sachs disease) are approvable; (b) attempts to produce "better" human beings are highly questionable. Who can say, or has the right to judge, what would be better?

Technological developments proceed so rapidly that we hardly have time to think about the ethical questions they raise. Nevertheless, the questions must be faced, if not sooner then later, all too often after damage has been done. American Protestants affirm the separation of church and state; they also have the freedom to express their views as part of the democratic processes by which issues are decided. Both religious and nonreligious moral judgements will be heard, and nowhere more obviously than in the field of medicine.

Part Six

Legal and Other Perspectives

Legal Aspects
of Procurement

Eric C. Sutton

Introduction

Laws are not created in a vacuum. They are shaped by numerous social forces that bring them into existence. The forces of scientific discovery and new technology often give birth to new laws designed to serve the needs and desires of an ever demanding society. In general, the birth of new laws does not go either unnoticed or unchallenged. The proponents of such new laws are often confronted with serious questions that challenge not only the efficacy of these laws but also their ethical foundation.

The emergence of transplantation technology is one such social force presenting unique ethical challenges. New laws are evolving. They spark serious ethical and moral debates that promise to grow more heated and complex with each new scientific discovery. What follows is a brief analysis of some of these laws and the ethical debates that surround their conception. This brief study is not meant to be exhaustive. The subject matter is growing at a pace no single text, or author for that matter, can hope to encompass completely. The field, therefore, is open for inquiring minds who are not deterred from traveling down roads where the traveler's concern for "doing the right thing" must meet and wrestle with the often harsh and haunting realities of the struggle for life over death.

From: *New Harvest* Ed.: C. D. Keyes
©1991 The Humana Press Inc., Clifton, NJ

Determination of Death

In the case of cadaveric donations of hearts, lungs, kidneys, and livers, the question of when a person is "really dead" must be answered before transplantation of these organs will legally be sanctioned. The emergence of transplantation of organs and tissues from the realm of the imagination into the bright lights of the hospital operating room has been a major force in bringing about a change in the legal definition of death.

It was not until respiration and heart function could be indefinitely maintained through the use of artificial means that society was forced to reexamine how death is defined both clinically and legally. From a legal perspective, how death is to be defined is decided by judicial determination i.e., case law) or legislative enactment (i.e., statutory law).

Case Law

Prior to development of the respirator and other machines that would maintain heart function and respiration for indeterminate amounts of time, state courts invariably defined death in terms of irreversible cessation of heart function and respiration or other similar language (*Baylor Law Review*, 1975). The need to reexamine this traditional medical and legal definition began in the hospital but moved into the courtroom. The case of *People* v. *Lyons* serves as an example of how such traditional court-made definitions can be altered through the clash of persons and ideas in unlikely settings (*People* v. *Lyons*, and analyzed *Baylor Law Review*, 1975).

In the Lyons case, a young man was shot, and after determination that his brain was no longer functioning, he was declared dead. His heart was then transplanted into another individual. After being charged with murder, the assailant raised the defense that removal of the heart caused the victim's death, basing his argument on traditional definitions of death as found in prior case law. The court, as a matter of law, rejected the defendant's arguments concerning the time of the victims demise and revised the traditional legal definition of death to include "irreversible cessation of brain function criteria" when based on "usual and customary medical standards of medical practice" (*People* v. *Lyons*, 1974). The California Supreme Court thus demonstrated its willingness to modify and modernize traditional legal definitions when faced with expert testimony as to generally accepted medical practice. Other court precedents have similarly brought outdated definitions of death in conformity with the medical community's revised viewpoints as to how death is to be defined (*Strachan* v. *John F. Kennedy Hospital et al.* and *People* v. *Eulio*).

Statutory Definitions of Death

Arguably, the courtroom, with its evidentiary and procedural constraints and limited access for public participation and comment, is not the best forum for debating complex legal, medical, and ethical questions that affect the health, welfare, and well being of the citizenry of a state. Courts, under the banner of "judicial restraint," are often hesitant to "make law" under circumstances where the judiciary believes that the legislature is a better judge of substantive public policy issues. In addition, the nature of how a case works its way through the maze of trial and appeal to finally become law often leaves unsettled and ambiguous important public policy issues that need to be confronted.

It can be argued that how death is to be defined is best decided by legislatures and not by courts. Death demands certainty in its legal definition. The Lyons case demonstrates this. So does the need to provide civil and criminal immunity for physicians who act in accordance with usual and customary medical standards used to define death. Consequently, state legislatures have wrestled with the complexities of how death is to be defined and provided a more modern definition of death.

Uniform Determination of Death Act

A majority of states have promulgated statutory definitions of death that include irreversible cessation of brain
function as a part of the definition (Capron, 1987). The desire to bring uniformity to the legal definition of death and
to eliminate uncertainty as to its definition has given birth
to the Uniform Determination of Death Act. This act de-
fines death as follows:

> An individual who has sustained either (1) irreversible ces-
> sation of circulatory and respiratory functions, or (2) irre-
> versible cessation of all functions of the entire brain,
> including the brain stem, is dead. A determination of death
> must be made in accordance with accepted medical stan-
> dards. (President's Commission, 1981)

The Task Force on Organ Transplantation has recommended that all states without brain death legislation adopt
the Uniform Determination of Death Act "to provide for
uniformity and to reduce all legal uncertainty." (Report of
Task Force, 1986) In addition, believing that legislation alone
would not relieve "misunderstandings" about brain death,
the task force recommended that model hospital policies
and hospital protocols be developed for determining death
based on irreversible cessation of brain function (report of
Task Force, 1986).

Support by the public of transplantation of organs and
tissues depends in large measure on the public's perception that these life-saving or life-enhancing procedures are
done with due respect for the sanctity of the donor's life
and that all appropriate medical measures are taken to
preserve and prolong the donor's life before any donation
will come about. The public's confidence in the mechanisms
used to identify and retrieve organs and tissues under "the
right circumstances" must be built on laws that clearly define
death and relieve any uncertainty as to when this occurs.

To do otherwise will lead only to doubt, confusion, and an undermining of support for organ and tissue donation.

Ownership of Organs and Tissues

The harvesting of organs and tissues for transplantation has been built around a framework of voluntary donation. Public education efforts with regard to the critical shortages of organs and tissues for transplantation (of one's body parts or those of a family member) invariably emphasize the altruistic nature of "giving the gift of life."

It has been, therefore, asserted that "any perception on the part of the public that transplantation unfairly benefits those outside the community, those who are wealthy enough to afford transplantation, or that it is undertaken primarily with an eye toward profit rather than therapy will severely imperil the moral foundations, and thus the efficacy of the system" (Caplan and Bayer, 1985). In keeping with this sentiment the National Organ Transplant Act makes it a federal felony punishable by imprisonment or fine to sell or buy organs if the sale affects interstate commerce (National Transplant Act of 1984). Additionally, the Uniform Anatomical Gift Act makes the sale of organs and tissues a crime punishable by imprisonment and fines (Uniform Anatomical Gift Act with 1987 Amendments).

The Task Force on Organ Transplantation suggested that, among other societal values, organ donation is based in part on the concept of "promoting a sense of community through acts of generosity" (Report of Task Force 1986). The National Organ Transplant Act, with its harsh penalties for the sale of organs, gives legal support for this ideal. This legislation supports the concept that donation is not done for profit. It has been asserted that, "These gifts of the body, ministering to the needs of strangers, connect us in our mutual quest to relieve suffering and to pursue our good, separately and together" (Murray, 1987).

Uniform Anatomical Gift Act*

Existing Law

Every state in the US has adopted the Uniform Anatomical Gift Act as law to legally facilitate donation of organs and tissues for medical purposes (UAGA, Original Model Act).** When first drafted and recommended for enactment by all states, transplantation of organs and tissues was in its infancy. The explosion of transplant technology has reinforced the importance of this legislation and also suggested the need for its refinement through statutory amendment.

The current "model" version of the UAGA allows an anatomical gift to be made by either an individual prior to his or her death or by a deceased's relative after death. An individual can make this donation by will or by another document that has been signed by the donor "in the presence of two witnesses who must sign the document in his, the donor's presence." The original model act also allows for donation of organs and tissues by next of kin upon death of a relative in order of priority stated in the act. The model act also provides that any person who acted in good faith in accordance with terms of the act is not subject to criminal prosecution or civil liability for damages.

Despite the objective to "eliminate uncertainty as to the applicable law" in respect to organ and tissue donation, the original UAGA has been criticized for "not producing a sufficient supply of organs to meet the current or projected demand for them" (Caplan and Bayer, 1985). A substantial

---

*All quotations in this section of the text are from the original version of the Uniform Anatomical Gift Act.

**The Uniform Anatomical Gift Act, like all proposed uniform laws drafted by the National Conference of Commissioners on Uniform State Laws, is a model act. A state legislature may, at its option, choose to adopt, ignore completely, or adopt with revisions such a proposed uniform set of laws.

revision of the act was drafted by the National Conference of Commissioners On Uniform State Laws in 1987 (UAGA, 1987). The proposed revision makes a number of significant changes in the earlier model act.

Proposed Revised Act in Comparison to Existing Laws*

Traditionally, despite language in the original model UAGA to the contrary, the medical community has been reluctant to harvest organs and tissues from a deceased without consent from appropriate relatives, even where the deceased has made his wish to donate known by proper execution of a will or other documents of donation prior to death. The 1987 proposed UAGA revision, according to the commentary accompanying the draft, strengthens the original intent of the act that consent of next of kin after death is not required where, prior to death, the deceased has executed the appropriate documents of donation. "An anatomical gift that is not revoked by the donor is irrevocable and does not require the consent or concurrence of any other person after death of the donor." The revision further sets out the only acceptable means for revocation or amendment of the gift.

Another significant change in the proposed revision is the granting of authority of the coroner, medical examiner, or local public health official to make a donation of *"any* [author's emphasis] part from a body within the coroner's [medical examiner's] custody, for transplant or therapeutic purposes" where there are no known objections from the decedent or other statutorily defined individuals. In some states, coroners or medical examiners already have statutory authority to remove eyes or corneal tissue, which has

---

*All quotations in this section of the text are from the Uniform Anatomical Gift Act with 1987 Amendments and the accompanying drafters comments.

led to "a significant increase in the supply of corneal tissue" in states where this is permitted. This proposed addition to the UAGA "is not limited to eyes or corneal tissue." Presumably, the hope of the drafters of the revision was to produce the same significant increases in solid organ donation that have been experienced with eye and corneal tissue donation.

Other important amendments or additions to the UAGA include adding a required request section (discussed in more detail below) and a section prohibiting the sale or purchase of an anatomical gift.

Whether all of the states adopt the revised model UAGA remains to be seen, but its submission to each state legislature for possible enactment will surely generate much debate.

Required Request Laws

State Required Request Laws

In the final report of the Task Force on Organ Transplantation, it is stated "that a major problem with the current voluntary system of organ donation is that families often are not informed of their option to donate organs and tissues after brain death is determined" (Task Force Report, 1986). The report went on to record the task force's recommendation that "all state legislatures formulate, introduce, and enact routine inquiry legislation" (Task Force Report, 1986). Such legislation would require statutorily designated hospital personnel to approach a deceased's next of kin, discuss the possibilities of organ and tissue donation, and inquire if they would like to make a donation for transplantation.

A majority of states have passed some form of required request legislation. There is a considerable amount of variance in existing state required request legislation and data is just now being accumulated to measure the effectiveness of these laws.

Federal Required Request Laws

The Task Force on Organ Transplantation also recommended that the Health Care Financing Administration require hospitals to have routine inquiry policies in order to participate in Medicare reimbursement programs (Task Force Report, 1986). In response, Congress mandated that all hospitals receiving Medicare and Medicaid reimbursement adopt routine inquiry/required request protocols (OBRA, 1986). This is an especially significant piece of legislation, given the dependence of hospitals on funding from this source.

Selected Ethical Issues

Sale and Purchase of Organs and Tissues

The Task Force on Organ Transplantation adopted the view that "organs are donated by individuals for the good of the public as a whole" (Task Force Report, 1986). Much of the literature designed to promote the virtues of organ and tissue donation uses altruistic language to encourage people to consider organ and tissue donation as giving "the *gift* of life." Any perception that organs and tissues are being bought and sold like any other commodity can seriously undermine donation of these already scarce human resources. Concern about this has prompted federal and state legislation prohibiting the sale of organs and tissues.

Presumed Consent vs Required Request

As stated above, hospitals must now establish protocols for identifying potential organ and tissue donors and approaching next of kin to inform them of their options to donate and to ascertain their willingness or unwillingness to do so. Other countries have adopted "presumed consent"; i.e., on determination of death, it is presumed that the deceased has consented to donation unless there is actual knowledge of an objection on the part of the deceased or their next of kin.

This approach has received some advocacy in the US (DeChesser, 1986). Significantly, state presumed consent or "removal in the absence of objection" statutes are common insofar as removal of corneal tissue and pituitary glands are concerned, and represent a "sharp contrast to other areas of organ and tissue procurement" (Overcast, 1987). Removal of corneal tissue is a much less invasive procedure than the removal of solid organs, and may account for the widespread adoption of presumed consent statutes for corneal tissue and not for other organs and tissues. The proposed amended model UAGA would allow medical examiners to follow this approach for all organs and tissues (UAGA, 1987).

The debate over required request vs the presumed consent approach to procurement of organs and tissues will probably continue for sometime on a state-by-state basis. The outcome of such debate will, in part, be determined by how successful the existing required request statutes and regulations are in producing an increase in the supply of organs and tissues and whether the process, as it is currently evolving, does not generate any controversies in the way families are being approached for consent. Required request legislation, which is a less aggressive approach to procurement of organs and tissues than presumed consent legislation, has met with criticism.

A recent criticism raises questions of clinical psychological, and social/economic "conflicts of interest" that have led the critics to conclude that "required request laws are inherently mistaken and must be substantially amended or repealed" (Martyn et al., 1987). One leading proponent of the required request legislation has attacked these criticisms as being made without first giving required request legislation an opportunity to work and diverting attention from the need to focus on the medical profession's failure to comply with these laws and better educate the public about the value in transplantation (Caplan and Bayer, 1985).

Whatever the merits of either argument, it is apparent that any legislation to increase organ and tissue donation is doomed to fail without the health care professionals' (physicians, nurses, hospital administrators, and other medical support personnel) support of organ donation and transplantation. The Task Force On Organ Transplantation concluded that the "failure of many health professionals to participate in the organ donation process will remain a major barrier to organ donation." (Task Force Report, 1986) This conclusion led to the task force's recommendation that medical and nursing school curricula incorporate education on organ and tissue donation and transplantation (Task Force Report, 1986).

Any such professional education program must have at its core an ethical framework that places emphasis on the value that organ donation plays in helping families cope with the loss of a loved one under unexpected and tragic circumstances and, at the same time, vests the medical profession with the notion that they are the guardians of important public resources that can lead to renewed or enhanced life for others. If our society is to maximize the retrieval of these resources, which even under the best of conditions will be in short supply, the medical profession must accept an ethical obligation to advocate retrieval of organs and tissues (while still respecting the rights of individuals to say no). This is a good stewardship of a public resource. If the Task Force on Organ Transplantation is right in categorizing physicians and surgeons of patients dying in hospitals as "gatekeepers" in "raising the issue of donation with families," then they must assume the ethical obligation of not obstructing the gateway (Task Force Report, 1986).

Anencephalic Newborn Donors

At the heart of transplantation is the proper identification of suitable donors. At least in the case of brain dead individuals, improved diagnostic techniques, professional and public education on the meaning of brain death, and well reasoned legislation should do much to alleviate fears that premature or improper screening and selection of potential donors for transplantation is occurring, and minimize embroiling this new technology in a maelstrom of controversy. However, the potential for controversy and heated debate exists in the area of use of anencephalic newborns as organ donors.

Approximately, 2,000 to 3,000 babies are born every year with a severe and fatal neurologic defect known as anencephaly (Capron, 1987). Children born with this condition invariably die shortly after birth. An infant that is born with anencephaly does not have a cerebral cortex but does have a functioning brainstem that "sustains and regulates a wide variety of bodily functions, including... spontaneous respiration" and most are not, therefore, brain dead under current definitions of brain death (Arras et al., 1988).

Despite the fact, donation of their organs for transplantation has been advocated (Anencephalic Organ Donation Committee, 1987). The advocates point out that such donations are not only of obvious benefit to the potential recipient, but also provide the donor's family with the feeling "something good might come out of a tragic situation" (Caplan, 1987).

They argue that anencephalic children are "functionally equivalent" to being brain dead or they lack the necessary prerequisite of "personhood" and so their body parts can be used (Harrison, 1986; Holzgreve, 1987). To open the door to donation from the anencephalic without legal obstacles, some proponents of such donations would amend brain death legislation to define the anencephalic as being legally dead and/or amend the *UAGA* to sanction such retrievals (Caplan, 1987).

The use of anencephalic infants as organ and tissue donors has met with serious criticism (Arras and Shinnar, 1988; Capron, 1987). Changing the legal definition of death to include the anencephalic infant is, one critic argues "a bad idea because either it will treat differently one group that is identical on the relevant criteria (the permanently comatose, who are dying and lack consciousness) or it will lead to a further revision in medical and legal standards under which the permanently comatose would also be regarded as dead although many of them can survive for years" (Capron, 1987). Critics also point out that "anencephalic infants are not dead but dying *persons* [my emphasis] and entitled to a full measure of dignity and respect" and treating them in a way that views them solely "as a means to benefit another constitutes a paradigmatic violation of moral law" (Arras and Shinnar, 1988). It has also been suggested that in the long-run, amendment of the *UAGA* to allow anencephalic donation would, with all the likely cumbersome safeguards that would have to be built into the law, in fact, undermine public confidence and support of transplantation (Capron, 1987).

Debate on this issue should continue to be heated as parents of potential organ transplant recipients and parents of anencephalic children, looking for some meaning to come out of the lives and deaths of their children, unite to force society to provide an answer to the difficult legal and ethical questions that this type of organ and tissue donation poses.

Vegetative-State Donor

It has been estimated that approximately 10,000 people in the US live in a persistent vegetative state and, hence, are incapable of thought and have no awareness of world and their place in it (Stein, 1987). Since they have a functioning brainstem, they can exist in this state for years. People in this state of existence breathe on their own without mechanical assistance. They do not, therefore, meet the clinical or legal criteria for determination that they are dead.

It has been suggested that "once consciousness is gone the person is lost" and "what remains is a mindless organism" (Stein, 1987). If one adopts this viewpoint, persons in this state of nonexistence might be considered suitable candidates for donation of their organs and tissues. This view has been vehemently attacked by some as being "grossly inappropriate" and "dangerous" (Stein, 1987).

Transplantation technology, therefore, has engendered a renewed debate over how "personhood" is to be defined and, once again, poses the ageless question of what the intrinsic value of a person's life is apart from any utilitarian value that the individual's life has for society as a whole. The debate in this area, obviously, has far broader implications for society than the narrow issue of transplantation. It raises the fundamental question of rationing access to medical care and other services that bring individuals together as a community in the first place.

Fetal Tissue Donation

One new areas of inquiry is that of transplantation of aborted fetal tissue. Fetal tissue transplants are currently being studied for treatment of a wide variety of disorders such as Parkinson's diseases and epilepsy (Bard, 1987).

The use of fetal tissue for either experimentation or for transplantation will obviously meet with traditional arguments for or against abortion. In addition, however, the prospect of using fetal tissue in transplants adds a possible commercial or "other value equation" to the abortion issue (Warren et al., 1978).

Aborted fetuses must now be viewed as having a possible life-saving or life-improving value to others. This places abortion within a new ethical framework. The sacrifice of a fetus now may lead to the improvement or extension of another life. Or, as one critique suggests:

Old questions must now be reexamined in light of new knowledge: What are our moral obligations (if any) to a fetus that has been or may be aborted? What are our obligations to the pregnant woman who undergoes abortion, whether spontaneously or electively? What are obligations to an individual whose health can be restored through transplantation? What are our obligations to the larger society, in which many may benefit through research support and therapeutic applications? (Mahowald et al., 1987)

The forces that oppose abortion as an elective procedure will certainly line up to oppose use of fetal tissue for transplantation as they will surely view with alarm any procedure that would cast abortions in the light of having any positive benefits for society.

If abortions continue to enjoy legal approval, advocates of the use of fetal tissue for transplantation will have to insure that rules resting on firm ethical principles are formulated and strictly adhered to in order to avoid transplantation from being dragged into the battle between the "Right to Life" and "Freedom of Choice" forces. Given the nature of the debate and the controversy that the debate has created, this will be no easy task. Transplantation in general has enjoyed the support of the public. Any perception that transplantation is an incentive or fuel for abortions is bound to erode the confidence of the public in transplantation. We have come too far to take a giant step backwards.

Conclusion

Society is being forced to reexamine how a national "community" such as ours, founded on principles of respect for the rights of individuals and providing the opportunity to enjoy the fruits of new technology, should handle the scarcity of human and other resources and the rationing that such a scarcity is bound to necessitate. We have a growing concern over how to deliver the best medical care to the most number of people without ignoring the unique needs of each individual.

Laws intended to maximize availability of these unique and precious resources must be constructed on a firm ethical foundation. To formulate laws without such a solid foundation can only serve to undermine society's goals in the delivery of these "gifts of life" to the most number of people in need. They can serve as a constructive model of how laws should be formulated to deal with the other improvements to the quality of life to which advanced technologies give birth. They can just as easily serve as an example of how a society, bent on fostering new technology, can lose its way when it constructs laws with little regard to the importance of using strong ethical building blocks.

Legal Aspects of Allocation

Eric C. Sutton

Introduction

Shortages in the supply of any resource inevitably lead to rationing and the development of policies designed to distribute those resources in an equitable manner. When the demand for any item far exceeds the supply, policies formulated to distribute those scarce resources must necessarily discriminate between potential recipients. The end result will be, therefore, that some will profit by those policies and receive the resource at the expense of others who must go without. For most Americans, the decision whether they are to be on the receiving end of a rationed resource is not a life or death decision. The policies governing the distribution of organs and tissues for transplantation does, however, present our society with such a life or death choice for many. The harsh reality is that demand for organs and tissues suitable for transplantation is increasing at a much greater pace than is the number of actual retrievals of these scarce human resources. Both federal and state laws have been enacted that are designed to promote and facilitate donation of organs and tissues for transplantation. However, these laws are still in their infancy and their impact on donation has not yet been felt or accurately measured.*

*The US Congress has enacted, for example, legislation requiring hospitals receiving Medicare and Medicaid funding to develop routine organ and tissue inquiry protocols (Omnibus Budget Reconciliation Act of 1986 [OBRA 86], Pub. L. 99-509). A majority of states have enacted

(continued)

From: *New Harvest* Ed.: C. D. Keyes
©1991 The Humana Press Inc., Clifton, NJ

The equitable distribution of these "gifts of life" is the subject of a continuing debate that has far reaching legal, political, economic, and ethical implications. How society addresses these questions and adopts policies to maximize the distribution of these scarce human resources has implications that extend beyond the narrow confines of a small branch of emerging medical technology. The issues raised by rationing of organs and tissues can reveal much about how this society or any society values the life of any individual or small group of individuals and how much that society is willing to spend of its national resources to benefit that small group. In a sense, what is being revealed by such a debate is society's conscience. Such a revelation may be disturbing, but is surely one from which we all can profit.

The Network for Distributing Organs for Transplantation

The transplantation of human organs into patients who are dying or whose health is seriously compromised has become, within a remarkably short period of time, a customarily accepted medical procedure for many diseases. The advances are the result of many factors, not the least of which is the refinement of surgical techniques and the use of new and improved immunosuppressive therapies. These improvements have not gone unnoticed by either the medical community or the public, and the number of people waiting for this form of treatment is at an all time high. The

(continued from previous page)

similar legislation (Evaluation of Methods Used by States to Expand the Number of Organ and Tissue Donors, Final Report Executive Summary, April 1988, prepared by Maximus, Inc. for the Health Resources and Services Administration, under HRSA Contract Number 240, 86*0048). In addition, the 96th Annual Conference of Commissioners of Uniform State Laws in August 1987 adopted a substantial revision of The Uniform Anatomical Gift Act of 1968 (Uniform Anatomical Gift Act with 1987 Amendments).

problems of those awaiting transplantation and those who have received a transplant have been well publicized and have become the subject of much State and Federal legislation. One of the most far reaching of these laws was the passage in 1984 of the National Organ Transplant Act (Pub. L. 98:507).

The National Organ Transplant Act is important because it led to the creation of the Task Force on Organ Transplantation (hereafter referred to as the "Task Force"), which was charged with the responsibility of conducting "comprehensive examinations of the medical, legal, ethical, economic, and social issues presented by human organ procurement and transplantation" (Sec. 101 of act).

The Task Force was given a broad mandate to conduct a multidisciplinary study of all problems and aspects of organ transplantation. Its multidisciplinary investigation led to the publication of a comprehensive report (hereafter referred to as the "Task Force Report") that addressed many legal and ethical concerns touching on, among many other issues, the issue of the equitable distribution of organs for transplantation. The Task Force Report represents an exhaustive study of transplantation in the US. It also builds a foundation for a comprehensive consensus on the development of a nationally solid organ transplant policy.

National Organ Procurement and Transplant Network

At the same time that the US Congress created the Task Force, it provided for the establishment of a National Organ Procurement Transplant Network, frequently referred to as the OPTN (Sec. 372 of act). As envisioned by Congress, the OPTN was to be established by contract between the federal government and "a private nonprofit entity which is not engaged in any activity unrelated to organ procurement." (Sec. 372 of act) The OPTN was given, among other responsibilities, the task of establishing a "national system, through the use of computers and in accordance with established medical criteria, to match organs and individuals included in the list, especially individuals whose immune systems makes it difficult for them to receive organs" (Sec. 372 of act). Policies adopted by the OPTN can mean the difference between life and death for many solid organ transplant candidates. It is significant to note that these highly critical issues of preference and priority are not being left solely to the transplant community to decide. The initial contract to establish the OPTN further provides that the Board of Directors may not be composed of more than 50% surgeons and physicians directly involved in transplantation.* In addition to public participation at the Board of Directors level, the OPTN solicits comments from the public as well as the transplant community when addressing major issues relating to equitable distribution of organs.** This trend toward more public participation in important legal and ethical questions cannot be ignored and may signal more openness to the public of what once was the domain of a small group of medical professionals.

Selected Issues Involving Distribution of Organs

Economic Rationing

Organ and tissue transplants generally remain an expensive form of medical therapy that benefits a relatively small portion of the general populace. Like other forms of treatment, transplantation must compete for funding from

*The United Network for Organ Sharing was awarded the initial contract to establish the OPTN by the Health Resources and Services Administration. The contract is subject to periodic review and cancellation.
**Author's personal knowledge as to policy-making process. Author was elected to serve on the Board of Director's of the OPTN in 1988 as a representative of the American Council on Transplantation.

both public and private sources. The concern over the rising cost of health care and the federal budget deficit has sparked a new round of debates concerning the need "to consider rationing of health care because it costs too much" (Besharov et al., 1987). In this environment, transplantation is a highly visible and easy target for rationing. In 1987, for instance, in the realm of publicly funded economic assistance, the Oregon state legislature voted to discontinue Medicaid support of organ transplants and elected to spend the money on prenatal care (Egan, 1988). The following reasons have been suggested as to why transplantation was, in this instance, singled out for removal of support through public financing (Welch and Larson, 1988):

1. Transplants are an expensive, highly visible form of treatment;
2. Only a small percentage of the overall population benefits from these treatments;
3. Transplants must compete with older, more established therapies; and
4. Current Medicaid legislation does not make coverage of transplants mandatory.

This decision is in stark contrast to previous national policies relating to public and private funding of kidney transplants and kidney dialysis treatment. The Task Force found that "virtually all kidney transplantation procedures are paid for by Medicare, Medicaid, or private insurance." (Task Force Report) The Oregon decision may, therefore, signal, as one commentator has suggested about health care in general, "a shift from clinical to financial control of ethical decisions" (Kolata, 1988).

Many private health insurance plans and policies allow employers and others to make a choice as to the extent of coverage contained in those policies or plans. Therefore, employers and others responsible for choosing the extent of coverage in any given plan or policy may likewise elect to treat transplantation as an expendable form of therapy.

Obviously, decisions of this nature generate a considerable amount of publicity and controversy, and spark renewed debates concerning the limits placed on the cost of health care (Kolata, 1988). The continuing debate in this area not only touches on economic and legal issues, but, by necessity, has an ethical undertone. Those who would substantially limit or altogether eliminate either public or private financing of transplantation therapy should consider these factors, among others:

1. There is a natural limit to the maximum number of solid organ transplants that can be performed in any given year since there is only a limited pool of potential donors (Task Force Report).
2. Private insurers are probably in the best position to spread the costs of these procedures among their insured, and can mitigate large expenditures through reinsurance (Task Force Report).
3. Patients who do not receive transplants still must be treated using expensive conventional therapies that fall within the usual and customary treatment coverage of both private and public insurance and health plans.
4. Transplant therapy is no less cost-effective than other therapies that have not been targeted for elimination or drastic reduction in economic support in the same way that transplantation has been targeted (Joyce et al., 1988 and Malcom et al., 1988).
5. Is it realistic and ethical to request organ donations on behalf of those who have either the public or private resources to pay for transplantation while denying access to those who are insured or under-insured and cannot afford to receive this form of treatment (Strom, 1988)?
6. What new discoveries or refinements in medicine in general, which inevitably spin off from new technology, will be sacrificed or delayed in their evolution should transplantation not be allowed to develop along with other better established therapies, and what is the cost in economic, social, political, ethical, and philosophical terms of losses or delays?

7. Is it fair that transplantation be singled out for treatment differently than any other form of therapy just because the law may allow for this, and because only a relatively small group of persons within the rest of society will be affected by such an approach (Sutton and Hunsicker, 1988)?
8. Should transplantation be made a scapegoat for society's failure to anticipate and institute appropriate controls over skyrocketing health care costs in general?
9. Should ability to pay ever be a criterion for access to critical, nonelective health care?

On the other hand, proponents of broad based public and private funding of transplantation must remember that, because the cost of health care in general has become a major public policy concern, proponents of expenditures on transplantation cannot expect to escape scrutiny nor the desire to limit its cost. Our resources have limits. The experience with federally funded kidney dialysis and transplant programs have necessarily made public policy-makers skeptical of any new advance in technology that bears a high price tag.

It would be a mistake for either side of the cost containment debate, however, to minimize the importance that ethical concerns will play in the resolution of the tremendous problems faced in this arena. In the final analysis, the decisions made are not just translatable in dollar signs. The quality and duration of the lives of many Americans will be unalterably affected by the policies that economic rationing of transplantation will establish. The American sense of altruism and compassion for those of us who fall prey to the forces of nature must not be sacrificed on the altar of fiscal responsibility.

Patient Selection Criteria

Before patients are enrolled in a transplant program, they routinely undergo evaluation processes to determine their suitability to receive transplants. The extent of the evaluation varies from center to center, but generally follows well established criteria that takes into consideration a wide variety of factors. Traditionally, many factors are considered: medical suitability and financial, social, and psychological concerns.

Medical Criteria

The major focus of any patient's admission into a transplant program is on the patient's medical suitability for transplantation. Suitability "rests on an assessment of two kinds of factors: medical indications for the procedure and medical contraindications" (The New York State Task Force, 1988).* The extent of the evaluation process varies from center to center and depends in large part on the type of organ or tissue that is being replaced. Given the scarcity of available organs and tissues for transplantation, a well defined program of evaluation is important to assure that organs and tissue are implanted in persons who really will benefit from the transplant.

Economic Concerns

When potential candidates have inadequate coverage or no coverage at all, they are usually required to deposit a substantial down payment for the proposed procedure. These deposits "may be as high as $100,000.00." Obviously, raising such staggering amounts to finance these procedures, despite public appeals for donations and the general willingness of the public to donate to these causes, creates a formidable roadblock, and, for some patients, has been

*This report is an excellent and comprehensive source of information on the procurement and distribution of organs and tissues and the author has referred to it frequently for assistance in preparation of this chapter.

tantamount to receiving a death sentence (Clark et al., 1988). The ethical questions raised deal with the value that society places on the life of any individual. "Am I [or in this context is society] my brother's keeper?"

Psychological and Societal Concerns

Psychological evaluations remain a routine part of a transplant candidates evaluation for transplantation (The New York State Task Force, 1988). They include:

(1) The patient's understanding of and commitment to transplantation; (2) The patient's ability to tolerate the emotional stresses associated with transplantation; and (3) The patient's willingness to comply with the demands of the rigorous post-operative treatment that will last throughout the remainder of his or her life (Jonsen, 1984).

Objections to these criteria include

1. The incorporation of questionable social and economic judgments about the candidate;
2. Inaccuracy in forecasting future compliance and behavior; and
3. The realization that familial circumstances and personal traits are often beyond any individual's control, and for ethical reasons, should not be used to preclude one from treatment designed to save that person's life (Childress, 1981; Ramsey, 1970; and Daniels, 1981).

Whether psychological evaluations are to be given more importance in the future if the demand for transplants grows and the supply of donated organs remains scarce remains to be seen. As tools for patient selection, however, they will no doubt remain controversial. For that reason, centers that rely on these tests, should make full disclosure to potential candidates as to the relative weight these evaluations play compared with other selection criteria. Should a patient be excluded as a candidate, he or she should be advised of the reasoning behind such exclusion and be given an opportunity to demonstrate why the conclusions drawn by such an evaluation are inaccurate or why they should not be used to bar access to the program.

Social Worth

In the past, the allocation of scarce medical resources or forms of treatment based on the individual value to society or as a reward for exemplary behavior has been rejected on numerous grounds (The New York State Task Force, 1988). The Task Force cautioned that patient selection criteria should reflect medical judgments "rather than judgments of social worth."

Age

Age has often been used as a criterion to screen candidates for admission into a transplant program (Task Force Report). The relevance of such a criterion is dependent on a medical judgment that increasing age decreases the likelihood for long-term transplant patient survival (Task Force Report). The "cutoff" age has tended to vary according to the type of organ or tissue transplanted. The danger of excluding patients based on age is the possible interjection of nonmedical judgments, such as judgments based on social worth, into the selection process. The Task Force warned, therefore, that centers should avoid "unwarranted discrimination against older patients" (Task Force Report).

Foreign Nationals

One of the more complex questions that the Task Force dealt with was the issue of admittance of nonimmigrant aliens to transplant programs. Because of the success of the US transplant programs, foreign nationals have sought treatment in this country for diseases and conditions where transplantation is the only viable alternative. In this country, transplantation of organs into a foreign national necessarily means that an American citizen was precluded from receiving an organ donated by another American citizen. The Task Force recommended that:

> ...non-immigrant aliens not comprise more than 10% of the total number of kidney transplant recipients at each transplant center until the Organ Procurement and Transplantation Network has had an opportunity to review the issue and that extrarenal organs should not be offered for transplantation to a non-immigrant alien unless it has been determined that no other suitable recipient can be found.

A minority of members of the Task Force dissented from this viewpoint and recommended that the same policy for extrarenal transplants recommended by majority be applied to kidney transplants also.

One of the first responsibilities of the OPTN after its creation was establishment of a national policy in this regard. In May of 1988, the OPTN adopted a policy with regards to transplantation of nonimmigrant aliens. The policy generally provides that patient selection "shall be based upon waiting time and on medical and scientific criteria that are publicly stated and fairly and uniformly applied." Political influence, favoritism, and discrimination based on economic advantage, sex, or race are to play no part in the process. All OPTN members are subject to audit and review of their policies in this regard and the OPTN Foreign Relations committee is charged with the responsibility of reviewing the activities of any transplant center whose "proportion of non-resident alien recipients for any solid organ transplant exceeds the 10% guidelines recommended by the National Task Force to determine the circumstances upon which this activity has continued" (Policy Statement, 1988).

The OPTN has, therefore, established guidelines that are designed to assure the public that no alien is being given any advantage over American citizens due to political influence or economic status.

The problem of transplanting foreign nationals is a difficult one. It raises questions about how we are to behave in relation to others who may not share our same language, style of life, or culture, but who, nonetheless, hunger for a long and productive life. The struggle to live and somehow find meaning in the struggle knows no international boundaries, yet the realities forced on us by a limited supply of organs and tissues for transplantation, combined with the fact that they are being procured within the confines of our borders, has forced us to narrow our perspective at a time when international boundaries seem less relevant.

Final Selection Criteria

Probability of Success

Among the more salient issues that the OPTN has addressed is whether persons in the gravest medical condition should be given preference over those who are healthier when deciding who is to receive an available organ for transplantation. It could be argued that the healthier a patient is at time of transplant the more favorable the outcome is likely to be, promising a more efficient and less wasteful use of a scarce resource. In practice, however, transplant surgeons have generally given priority to patients in urgent need and patients for whom compatible organ matches are difficult to obtain (The New York State Task Force Report, 1988). The OPTN's current position is to assign a higher priority to hard-to-match patients and patients in urgent need.*

Retransplantation

A person who receives an organ transplantat must live with the knowledge that the graft may fail. Unless there are medical reasons to the contrary, the previously transplanted

*See the second note on p. 244.

patient becomes a candidate to receive another transplant. Therefore, previously transplanted patients compete with others who have not yet received a transplant. A question arises, therefore, as to whether a previously transplanted patient should receive preference over a patient who has not yet received a transplant.

Some would argue that a person should be limited to a set number of transplants because the scarcity of organs dictates that as many people as possible be given a chance to receive a transplant and that retransplantation seriously limits that possibility. Others argue that a substantial "investment" has been made in preserving already transplanted patients lives and (re)transplantation is an acceptable way of protecting that investment. It has also been urged that to "deny the patient another transplant would constitute abandonment" (Task Force Report). The Task Force took no specific position with regard to this issue. With respect to nonrenal solid organ transplants, the OPTN has adopted a policy that gives (re)transplant patients higher priority over candidates who are not in urgent need.*

Issues concerning the priorities to be established in distributing transplantable organs often involve competing sets of values, each of them founded on well established ethical and legal principles. This creates a tension that makes policy formulation extremely difficult and fraught with conflict. Perhaps no better guidelines can be established or improved on in this area than the following Task Force recommendations: (1) "The Task Force recommends that selection of patients both for waiting lists and allocation of organs be based on medical criteria that are publicly stated and fairly applied. The Task Force also recommends that the criteria be developed by a broadly representative group that will take into account both need and probability of success. Patients otherwise equally qualified should be based upon length of time on the waiting list." (2) "The Task Force

*See the second note on p. 244.

recommends that selection of patients for transplantation not be subject to favoritism, discrimination on the basis of race or sex, or ability to pay."

Whatever the specifics of those policies are, they should not be so rigid as to preclude revision and refinement. They should always be formulated with the understanding that lives are saved or lost, enhanced or diminished in the process. The responsibility in this area is awesome, but cannot be shied away from. To do so would only serve to further compound the uncertainty that already dominates the lives of those who await transplant and produce doubt in the minds of those who are confronted with the decision to donate. Given the scarcity of these precious resources and the good that can be accomplished through donation, such uncertainty is unacceptable.

Conclusion

The technology that has led to the development of organ and tissue transplantation has made a dramatic impact on the lives of many who would be doomed to certain death or whose lives would be severely limited without these remarkable medical advances. These procedures have sparked not only wonder, but also force us to reevaluate how we are to measure the worth of the individual against the countless demands that our citizens place on the political, social, economic, and our other societal systems and institutions. How we struggle with and resolve these complex issues can reveal a great deal about the fabric of this society and the values on which it is founded. There are many lessons to be learned from this inquiry for those who are not afraid to confront questions that, for some, have life and death implications, not the least of which is how far a society is willing to go to save the lives of those who, through no fault of there own, must await the call to receive the "gift of life."

A Word
from the Other Players

George L. W. Werner

Introduction

After the capture of Ticonderoga, Ethan Allen, leader of the Green Mountain Boys, attended a Thanksgiving service at the Old First Church in Bennington, Vermont. During the long prayer, in which the Rev. Mr. Dewey was giving all the credit for the victory to the Lord, Allen interrupted, saying, "Parson Dewey, Parson Dewey...please mention to the Lord about my being there."

As a community leader given responsibility for oversight of organ transplantations at three hospitals, I have heard repeatedly from patients, their families, and the community about the need to remind the medical professionals that, like Ethan Allen, they too are part of the battle. Each has a role in transplantation and we overlook that role at our peril. For almost five years, I have been deeply involved with the Pittsburgh Transplant Program. I have met regularly with the key players and have visited patients, families, and support groups. I am also a member of the board of Transplant Recipients International Organization, which offers ongoing support to those who have received grafts. In addition to these, I have given numerous speeches to academic and other organizations that emphasize the need for donation.

From: *New Harvest* Ed.: C. D. Keyes
©1991 The Humana Press Inc., Clifton, NJ

The Community

The United States of America, a land of liberty and unequaled resources, is severely threatened by a crisis in medical care. More than one-third of her citizens cannot afford adequate medical insurance, and more than one-sixth have no coverage at all. Although it is the home of outstanding research, state-of-the-art equipment, millions, even billions of dollars of grants for study and experimentation, the US has an abominable record in infant mortality, higher even than several third world countries. Towering hospital buildings with expensive scanners and ultrasound and lasers look down on areas where poor prenatal care and malnutrition in children are reaching epidemic proportions.

The issue is one of distributive justice. How should we allocate a resource when it is in short supply? Whose need should be given priority? Who has the first call on a tranplantable heart or liver? On a larger scale, should we spend 150,000–500,000 dollars on a single transplant operation when as many as 500 children might be provided with basic health insurance for the same amount of money? Should we tie up the hospital's intensive care units, top surgeons, laboratories, and other scarce and expensive resources so that a handful of people can trade a terminal illness for a lifetime of medical care?

On the other hand, should we stop these new and experimental procedures that may lead us to a new understanding of medicine and surgery and that could, in the long run, save multitudes? Should we allow hundreds of people to sacrifice their last chance at life, including many who have much yet to give, simply to get "more bang for the buck," to serve according to numbers rather than the intensity of need? Distributive justice is never a problem when there is an unlimited supply, but it is clear that the procurement of human organs is not keeping up with the demand. Short of using baboons or other animals, or developing artificial organs, distributive justice is the question.

The whole community has a stake in such matters, and thus, has a say. These are not matters that can be delegated exclusively to the medical professionals. Anyone who accepts responsibility for making decisions and establishing policies must give serious consideration to the ethical concept of distributive justice. To understand this concept better, Beauchamp and Childress (1983) suggest these principles:

1. To each person an equal share;
2. To each person according to individual need;
3. To each person according to individual effort;
4. To each person according to societal contribution; and
5. To each person according to merit.

How might such principles be applied to transplant operations? Which patient shall receive the scarce viable organ (in contemporary medical ethics, this is known as a problem of "microallocation")? Should all be given an equal share? Actually, in this context, we assume the second principle as well, which is "need." Patients are not nominated for organ transplants unless their need is acute. Equality then comes to mean equal consideration for equal need. Since the need is determined by medical standards, other considerations, such as the ability to pay or social status, should not count. A dishwasher should have the same chance as a corporate executive.

Since there now exists an insufficient supply of organs to meet those cases that demonstrate acute need, hard choices must be made. A traditional concept for such choices is triage. Usually, triage is the division of waiting patients into three categories: those to be treated immediately, those who can wait for treatment, and those who will be abandoned and probably die. During wartime, should a soldier with a minor wound be treated first so he can be sent back to fight again while a more grievously wounded soldier must wait?

Triage is never perfectly clean or simple. In the transplant field, some wish to give preference to those with a better prognosis. Typical comments are "We shouldn't waste precious resources," or "We need to get more bang for the buck." Too many comatose patients have made complete recoveries after transplant surgery to give support to such reasoning so far.

Even more troublesome is the fact that triage seems to bring into consideration the relative value of the persons for whom life and death choices must be made. Is it better to save the nineteen-year-old high school dropout or the elderly scientist on the verge of a major breakthrough in medicine? Should we choose the senator or an average housewife with minor children? It is at this point that other principles such as societal contribution and merit come into play, and challenge the principle of need. Unfortunately, there are usually tensions and conflicts between principles when they are applied to particular cases.

The Committee on Oversight of Organ Transplantation (COOT) was established in Pittsburgh in 1985 to represent the community's interests in organ transplantation. Though it includes members of the medical community, it is basically a lay committee. COOT met regularly with surgeons and other transplant personnel, hospital officials, and administrators of the Pittsburgh Transplant Organ Procurement Foundation to discuss an equitable system for allocating organs. Concerns such as time on the waiting list, medical urgency, and proximity to the center were deemed as important as antigens and antibodies. Those meeting reasoned that the community's perception of fairness in allocation was vital to support for organ transplantation. It was argued that procurement relies strongly on the people's trust. If there are increasing numbers of organs to be harvested, there must be the understanding that the rich would not be favored over the poor, nor the influential over the disenfranchised.

The Pittsburgh criteria were developed into a point system that could be computerized. The system was then adopted by the United Network on Organ Sharing for the US and Canada. The system is not carved in stone, and was amended in the first two years. There is still strong disagreement about the priority assigned to medical urgency, since some argue that it would be better to transplant patients before they reached intensive care or coma. Nevertheless, there is a clear commitment to equitable treatment for all under rules that can be readily monitored.

If scarcity is the root of allocation controversies, then increasing the pool of harvested organs is a basic objective. Unfortunately, efforts such as state laws requiring a request or campaigns to have people sign donor cards have not improved the situation so far. It appears that procurement professionals are far more effective in gaining a family's permission to harvest vital organs than are hospital personnel assigned to comply with the law.

The average nurse or social worker is trying to cope with the family's grief and not with what some see as the ghoulish responsibility to get the kidney or heart. Even when a family seems willing to give permission, if even one member of the family protests, the hospital is often too afraid of potential litigation to use those needed organs. The conflict between individual liberty and social responsibility needs clear legislation and appropriate criteria for procurement and allocation. We also need a massive effort to educate the general public as to both the need and desirability of donating these life-extending organs.

So far we have dealt with triage on the level of "microallocation." It can be extended to questions of "macroallocation" such as (a) how much of our resources should be given to transplants as compared to other medical needs, and (b) how much should be given to medicine and health care overall in relation to other social needs. Unfortunately, microallocation issues get more attention from the media and in the public consciousness because they are more individual and dramatic. A program offering five thousand children proper nutrition and health care does not make the same impact as the story of a cute little child waiting for a transplant in a race with death.

Assuming organ transplantation is a gift to the community, how shall it be a fair and just gift? The answer is the opening of an ongoing dialogue between the experts (the surgeons, the nurses, the coordinators, and others with skill, experience, and certification) and the lay representatives of the public.

In a century when the Nazis taught us that even those who had taken the Hippocratic Oath were capable of committing horrible atrocities in the name of experimentation and progress, it is vital to keep the community as part of the decision-making process. Lay people may not have the answers, but they do have important questions, and those questions are just as essential as those the researchers and medical personnel are trying to resolve.

The Patient

There is a saying in pediatric transplantation that only the toughest parents get through the system to obtain an organ for their child. Though exaggerated, it contains an important truth; it is a long and difficult process for the transplant patient to receive a life-saving graft, partly because of differing perspectives of doctors and patients.

Communication between the physician and the patient is vital, yet it is endangered by different agendas. Certainly, both agree on the quest to save the patient's life. Equally certain, physicians are usually altruistic and conceive of themselves as battling death or even the loss of quality-of-life. However, there are other driving forces for the physician who is part of a major breakthrough in health care.

When should the surgeon surrender on an individual case? Is there a point where the patient is too sick to regain quality-of-life? How many transplants are the limit for one patient before the public is outraged? The surgeons will tell us that when they have removed an organ vital to life, they feel responsible to keep trying because the alternative is certain death. There is no turning back. Therefore, the surgeons want each case to be determined on its own merits rather than some institutionally imposed general rule.

Top surgeons are, by definition, competitive. The articles one reads in major medical journals are not just for the purpose of information or challenge, but also for establishing one's credentials and, in some cases, for staking claims of original work. A major transplant surgeon in Pittsburgh is the point man in the development and testing of a new immunosuppressive drug. The early results have been so startling that he considers it a tragedy that he may not use it on every patient. Yet his hospital's Internal Review Board and some others in the wider community are cautious, wary of other wonder drugs, such as thalidomide, which turned out to have devastating side effects that far outweighed the drug's benefit.

The issue for us is not who is right, but to see this as the kind of pressure exerted on the physician/patient relationship. The top men in any new field are anxious to push the parameters, to expand the envelope of their findings. Yet the patient also has an agenda, which is usually to be given the best chance for life at the least cost, physically, emotionally, and, yes, financially. Obviously, these two agendas do not always coincide.

At a parents' support group meeting, many complaints were made concerning requests to try some new and dramatic procedure or medication that demanded an instant answer. Patients and their responsible family members are in a new world. They may be decisive and quick in their own vocations, but now they are in a place where someone they love may die because of their choice, and they need both information and time to make such decisions.

The physician who deals with such matters routinely needs to be sensitive to these complaints. Yet another popular saying is that the surgeon believes in saving the patient now and explaining later. Invariably, though, the physician is at the end of a seventeen-hour-day, and the nurse coordinators have too many patients to handle to spend sufficient time with one case. To oversimplify, the family is concerned with the life of a loved one, the medical personnel with defeating their old and continuous enemy of death, and conflict necessarily must result.

Joyce Willig was diagnosed in late 1971 with what her doctors called a "slowly progressing, ultimately fatal liver condition with no known cause and no known cure." For eleven years, the team at Yale-New Haven Hospital kept her alive until cyclosporine was available as an immunosuppressive drug. Mrs. Willig received her first liver on February 7, 1982, and when that failed, her second graft was on February 27.

Her recuperation took almost a year, and although she describes it as agonizing and traumatic, she enthusiastically supports liver transplantation and is thankful for the team that gave her extended quality-of-life. Willig says, "I lost hearing, speech, vocal cords. I had a lung collapse. I was put on a ventilator and was terrified." Patients and families are involved in the most traumatic and emotion-filled period of their lives and need to be provided with both a better understanding and a clearer picture, and to be partners in the transplant process.

Joyce Willig believes that the many seminars and presentations on transplantation need testimony from successful patients and from donor families. "We are all a part of this technology." Such seminars, though, seem to concentrate only on the medical and the technical aspects, and rarely, if ever, on concerns of recipients and donor families. Joyce sees this as a missing piece of the puzzle that results in unnecessary insensitivity to and ignorance of patient's needs as well as a lack of awareness as to the recipient's personal organ program and viable role to society.

The insensitivity of medical personnel was also an issue for Margaret Joseph. A registered nurse herself with extensive hospital experience, Mrs. Joseph and her husband George went through two tortuous years trying to get a kidney transplant for their infant daughter, Kristine. From the time of Kristine's birth, the hospital continually informed the parents that it would be their choice to intervene medically or to let the baby die. The Josephs saw Kristine "bouncing around the bed and too happy and full of life" to give up without a fight. If Kristine was willing to battle, then they would be too.

But the battle was made more difficult by physicians and nurses who would march into the room and give orders with no consideration of the parents. Margaret Joseph relates some incidents. "A resident thought it necessary to do a tracheotomy but I had been with Kristine constantly and disagreed, so I refused permission. The doctor then asked me, 'Don t you love her?'" Another time, when there was a genuine threat of sciatic nerve damage, a physician insisted on putting an arterial line in her leg. Again, the mother disagreed. "We parents are more traumatized than the patient."

More than most people, Mrs. Joseph understands that she was not to be the final medical word on her daughter's case, but seeing strangers wander in, read a chart and give orders without any dialogue or history with the family was frightening and frustrating. In addition, before the transplant, there were constant disagreements, confrontations, and debate between the parents and the pediatric physicians concerning Kristine's program. Now that the transplant has been successful, all of the physicians have become positive supporters of the progress Kristine has made.

When the baby finally received her father's kidney, recovery was swift. It was everything they had hoped for. "She's here, she's growing, she's wonderful...she keeps us laughing all the time. Seeing her feel good and breathe like any other kid, we know that miracles do happen."

Like Joyce Willig, the Josephs are very positive about transplantation, but they want to see more understanding of the patients and the donor families. On the day that Kristine was to come home, Margaret found out that a floor nurse had indicated to the attending physician that Kristine's mother had not learned to do medications yet, and therefore, Kristine could not be released. The nurse had never spoken to Margaret about this, nor did she know that Margaret was also a nurse and certainly understood medications. The very gentle, attractive, caring mother of a transplanted daughter told the floor nurse, "I'll rip your face off."

This is not intended to be a blind and general attack on medical personnel involved in transplantation. Rather, it is a reminder of the tremendous pressure and anguish of patient and loved ones. I hear repeatedly of "the gods in white" syndrome, the perception that these arrogant, skilled miracle workers are too important to be bothered by patients and families. I am aware of the fear of family members that, if they complain, it would somehow damage their loved one's chances for an available organ.

This is not what I observe when I listen to the concern and dedication of surgeons and support teams as they try to solve problems of a new and compelling breakthrough in medical science. Impugning motives and character assassinations are not called for. The problem is simply one of human limitations. Medical specialists will invariably view cases from one perspective, patients and families from another. We must simply try to assure that each will be given due consideration.

The Donor and Donor's Family

We must also consider donor families. The pressure can be overwhelming on those going through the emotional crisis of seeing someone, who is extremely important to their own life-being, dying. They must retain the right to say no and to make an actual choice. We also need to step up efforts of community education so that the thinking and deciding is basically done before the choice arises.

If this is true of cadaveric donors, it is even more important to consider the rights of living donors. For many years, kidneys from parents or siblings have saved lives. More recently, surgeons are experimenting with a graft of part of the liver from living donors. The liver regenerates for both donor and patient and would, according to some, supply one answer to the shortage.

Living related donors take a large risk. They face major surgery. If their remaining kidney is diseased later in life, they have no backup and will find themselves in the same position as the loved one they helped originally. There are also strong emotional issues, such as the parent whose organ is not used for the child, becoming a second class citizen to the donating spouse. Guilt should never be the leverage to find new organs.

Donor families also need excellent communication from the professionals. Often mentioned is the haunting feeling that perhaps more might have been done to save the loved one if his or her organs had not been needed for others. Even the slightest hint of impatience to begin the harvesting may raise the fears that the family has failed to protect its loved one sufficiently.

Although there are many donor families who feel that their gift has salvaged something important, others wonder, "Did we do the right thing?" "Was this what he really would have wanted?" This ambiguity requires the best trained personnel to be available to potential donor families at the crucial moment. If we spend precious resources and dollars on the technical aspects of transplantation, we need also to make available quality care for the emotional needs of the patients and those responsible for the organ donations.

PART SEVEN

Conclusion

Reaping the Benefits

C. Don Keyes

Introduction

Transplant surgery and related technologies are motivated by respect for life and have produced beneficial results. At the same time, they reap these benefits at the expense of conflict and push values related to them to their limits. Transplantation challenges limits by confronting us with ethical problems and tensions. If we respect life as given, how far have we the right to change it? We have even blurred the distinction between life and death through our ability to keep some life functions going although others have ceased. We teeter on the edges of life and death, and at the limits of right and wrong. There is also something to be gained through this uncertainty, though. As we are challenged to wrestle with the issues, we also have an opportunity to rethink our ethical ideas, including the basic concept of respect for life.

It is important to respect limits; some things we *can* do *should not* be done. It is also important to see what limits can legitimately be stretched or removed. Earlier chapters have examined and summarized elements of these and other ethical issues related to transplantation. Three main arguments are basic to the philosophical chapters and articles that unite this book:

1. Deontology and consequentialism are not irreconcilably different methods, for they have a common foundation in respect for life. Even though the two methods frequently

From: *New Harvest* Ed.: C. D. Keyes
©1991 The Humana Press Inc., Clifton, NJ

conflict with one another, they converge at other times. The complexity of transplantation requires the simultaneous use of both methods. The new reproductive technologies, especially gene transfer into human embryos, provide an unusually clear model of the elasticity of the two methods.

2. Many ethical concerns of transplantation are crises of self-identity. These include

    a. genetic self-identity in gene transfer;

    b. bodily self-identity in donation and reception and the symbolism of body parts; and

    c. neurological self-identity in

        i. brain death,

        ii. fetal brain function and the question of when human life begins, and

        iii. brain tissue grafting as it potentially affects the recipient.

3. The brain-mind problem is the core of the foregoing self-identity issues. Respect for life is reconstructed by working through, not going around, that problem. The monistic reduction of mind to brain, which appears to be required by scientific evidence, is the ground on which these three foundations of respect for life are built:

    a. The degree of monistic reduction is limited by criticizing "eliminative reductionism," which does not allow reference to mind. Even though mental states such as consciousness, intentionality, and so on are identical with the neurobiological processes that cause them, such states still have validity in their own right. The significance of an event is not reducible to its cause.

    b. Solipsism, the belief that only the self exists, is not credible. The interchangeability of body parts, the gift relation, and other phenomena related to transplantation vividly attest to the reality of other selves.

    c. The importance of use is limited by renouncing what the second chapter calls the illusion of entitlement to use the other." As the quotation from Kant (1785) at the end of that chapter claims, other human beings are ends in themselves, not merely "means to be arbitrarily used by this or that will."

This final chapter continues the discussion of limiting the importance of use. Therefore, it focuses on a single characteristic of the good that ought to guide public policy as well as decisions in the donor/recipient and physician/patient relations. That characteristic is respect for life and its irreducibility to self-interest.

Limiting the Importance of Use

To respect life is to shift interest from its utility to its intrinsic value. We are not entitled to use others as means that serve self-interest. Expressed positively, we value others as ends in themselves. In terms of the Hippocratic tradition, treating persons as ends translates into doing what is beneficial to patients and not harming them. Physicians benefit their patients by preserving and restoring the biological mechanism that makes their human self-identity possible. Of course, this identity includes the "responsive" and "expressive" capabilities mentioned in the second chapter, not merely survival. The desire to preserve life at all cost must be limited by respecting the right of living human beings to respond to their world and express themselves in it. Furthermore, avoiding harm means not dehumanizing patients and those they love. Not harming patients requires physicians to obtain informed consent from them. Patients have dignity and rights. Physicians must never display arrogance, but only sympathetic understanding, in dealing with patients. I still feel unresolved anger about the rude way a physician told me, when I was a teenager, that my father (himself a physician) was dying. I had accepted his death as the only solution to his suffering, but his physician's manner permanently damaged the acceptance that already existed.

The physician's ability to treat patients as ends and not means and to respect their humanity depends in a large part on the social and political order at the base of medical practice. According to Aristotle (384–322 BC), it is "nobler and more divine" to attain the good "for a nation and for states" than for an individual person. This should be interpreted to mean that the well being of individuals is especially affected by a particular product of politics, namely legislation. There is conflict about which model of the good ought to guide political answers to questions in which the whole society has a stake. Every society also has an economic order, and in our capitalist society one model of the good predominates. Plough (1986) claims that belief in the "sanctity of private enterprise" causes the "market metaphor" to be uncritically applied to transplantation or the treatment of disease. Belief in free enterprise and the profit motive is seldom self-critical. It tends to be a self-certifying reductionistic ideology that views institutions and activities merely in the light of the economic base that makes them possible and evaluates everything in that light alone. Hospitals and universities, to give two chief examples, are seen simply as businesses in every sense. This breeds a managerial elitism that sometimes seems more devoted to economic success (or the appearance of it) than to serving its patients' (or students') needs. Since that ideology is incapable of criticizing itself, public policy must challenge its egotism and limit its power.

Plough (1986) describes how at least one corporation somehow got copies of a manuscript he was going to publish and resisted disclosures it contained. He explains how the senior vice president of a health care corporation sent him a thinly veiled threat by "special delivery certified mail requiring a return receipt." In this excerpt, the threat is clear.

I am deeply concerned about your analytic method and submit that it presents a highly distorted image of the facts...I have therefore been advised by council to review my discussion with you in writing...This is not an issue involving freedom of the press or public expression of opinion...Please be advised that we intend to submit your study for detailed

statistical consultation, evaluation, and review and will seek
the appropriate remedies if our perceptions are confirmed.

I must admit that I ran into no conflict of that kind while
writing this book; yet I sensed potential resistance when I
was indirectly asked, "What is the actuarial value of a book
on transplantation ethics?" When I mentioned problems
like this to an attorney, he said during time for which he
charged me, "Cash is king, and you're learning it fast."

Models of legislation based on respect for life challenge market mentality models in which there is excessive
profit taking at the expense of human misery and the right
to health care. Respect for life requires a change from seeing
others only as means of personal gain to seeing them as
ends in themselves. According to Fromm (1956), "respect" is

> ...not fear and awe; it denotes, in accordance with the root
> of the word (*respicere* = to look at), the ability to see a person
> as he is, to be aware of his unique individuality. Respect
> means the concern that the other person should grow and
> unfold as he is.

Even worse, when we treat others simply as means to
our ends, we damage our own humanity. We are essentially
social beings. Social responsibility challenges the market
mentality's reductionism. Claiming that some other model
besides the market metaphor ought to guide policy decisions
is not an argument against the free enterprise system as
such. To challenge the limits of something does not mean
that it is invalid within its limits. However, the unique
quality at which all medical practice, including transplantation, aims cannot be stated entirely within the limits of
the market metaphor. In other words, healing is not purely
and simply a commodity. Transplantation's challenge of
the market metaphor calls attention to the overplus of
meaning that healing has, which cannot be stated entirely
in terms of business transactions. This challenge extends to
(1) the nature of the relation between donors and recipi-

ents, (2) the issue of the social responsibility regarding procurement and allocation of transplantable materials, and
(3) the relation of physicians to their patients and the
community.

Donor/Recipient Relation

Conflicts in the process of donation and reception push
the market metaphor beyond its limit. This process and the
"miracle of its gifts," to use Werner's language, join recipients
and donors in a unique bond that can include conflicting feelings. Some of these stem from contradictory struggles and the
recipient's possible identification with the donor.

Recipients sometimes experience mental states like
those described earlier. One is the "enlarged body image"
that some recipients experience, according to accounts by
Castlenuovo-Tedesco (1973,1978,1981). The interchangeability of organs reveals the flexibility of the body's composition, thereby refuting solipsism without intending to do
so. Body parts can symbolically "preserve traces of their
former identity." Recipients sometimes feel that these are
transferred from "one self to the other." Recipients also
sometimes feel emotional mixtures of elation, excitement,
and concern for the donor's health. There can be mixed
guilt and gratitude as well as ambivalence about the right
to have the life-saving organ. Fox (1978), cited in "Body and
Self-Identity" of this volume, refers to the nonreciprocal
nature of this gift of life. The recipient feels that the graft is
an inherently priceless gift for which it is impossible to
repay the donor.

The process of reception and donation is not a business
transaction. The bond of life and gift relation, as the recipient experiences it, cannot be reduced to utility alone.
Something similar is true of donation; the gift cannot be
required but it still ought to be offered. The altruism of
donors' families is revealed even amidst conflicts and uncertainties like those Werner describes.

Social Responsibility

Respect for life conflicts with policies controlled exclusively by the market metaphor. Procurement and allocation become more complex as a result of this conflict. The social responsibilities of the medical profession are especially critical in the ethical issues raised by the procurement and allocation of transplantable organs (tissues, etc.). The same is true for the allocation of medical and economic resources for transplant purposes. Procurement and allocation have been discussed in various parts of this book. Although no relatively brief summary can do justice to all the questions involved, at least two general observations can be made.

First, physicians have a special responsibility for strictly medical judgments about scientifically valid research and reasonably effective treatment, but they also have a responsibility to the larger society. As Werner says, the public has a right to ask whether the policies followed are fair, whether they represent "distributive justice." In regard to procurement, especially from living donors, there are problems about free and informed consent. For example, it is not enough to obtain a prospective donor's signature on a consent form. There are pressures on such persons to give consent, especially if they are related to the recipient. Nelson and Rohricht (1984) cite Henry K. Beecher as claiming that 40% of donors would refuse to donate if they were assured that their refusal would not be made known. According to Veatch (1977), there is the even more difficult question of whether a transplant operation should also include some benefit to the donor, or whether the donated organ should be considered a sacrificial gift. What is fair? What is just?

Second, in responding to such questions, it is important that the general public have a voice. This is in recognition that nonmedical persons have both an interest and a responsibility in these matters. The joint responsibilities of the medical profession and the "body politic" emerge even more sharply when it comes to questions about how much of our resources should be devoted to transplant operations in comparison with other medical and health needs of our society. Among other things, this raises questions about what we owe to the 37 million (more or less) persons in the US who have no coverage for ordinary medical care. In the end, this has to be decided by the society as a whole, but medical professionals surely have a special responsibility to consider general social needs. They are also citizens. Further, their professional practice is both honored and sustained economically by the society at large.

There are no easy or clear-cut answers to these questions. What is humane? What is just or fair? What protects the intrinsic value of human lives? Procurement affects the integrity of individual donors. The intrusion of a "foreign" organ also affects the integrity of recipients. Allocation raises questions about our mutual responsibility to use our resources in the best way to meet the most urgent human needs. In these essays, such questions have been addressed and some ethical judgments offered. The public debate over approvable policies will continue and will be altered by further developments in biomedical technology as well as by changes in our economic resources. What policies this society adopts will inevitably reflect some compromises between what seems to some of us to be ethically right and what may emerge as politically feasible.

Physician, Patient, and Community

Physicians undergo conflict when they make decisions that affect the life or death of a patient. Furthermore, their commitment to an individual patient sometimes conflicts with the social demand for distributive justice. At the same time, the ethical values of healing often conflict with one another. These and other tensions that spring from healing cannot be explained as if they were merely various species of self-interest contradicting one another. The significance of healing comes from its intended effect, namely the advantage of the patient. Plato (c 428–c 348 BC) attributes a statement to Socrates that calls attention to the nature of the physician's desire to heal: "Then, isn't it the case that the doctor, insofar as he is a doctor, considers or command not the doctor's advantage, but that of the sick man?" Transplantation shows in an unusually clear way how the professional conflicts with which physicians have to live push beyond the limits of their own advantage. No attempt is made here to explain all the ethical issues that have this effect. One particular issue, however, illustrates the limits in question with unusual clarity. The issue is retransplantation. If a transplant fails, should the same patient receive another transplant? Foundational arguments both for and against retransplantation reveal in a succinct and equally extreme way how physicians and others affected by transplantation wrestle with ethical limits. Retransplantation discloses a tension that belongs to a wide range of other problems of transplant surgery, namely the good in conflict with itself.

Arguments For Retransplantation

Deontological Thesis

The Hippocratic principle of beneficence and avoidance of harm requires surgeons to fulfill their responsibility to individual patients. When a surgeon removes a vital organ, that action makes it impossible for the patient to live without a replacement. Therefore, if the patient is competent to give informed consent, the surgeon has a duty to attempt to retransplant. This responsibility to the individual patient is absolute and takes priority over potential claims in favor of distributive justice.

Consequentialist Thesis

If the surgeon does not retransplant, many will be harmed directly. Abandoning the patient will also create a bad reputation for the medical profession and this will indirectly harm other patients.

Arguments Against Retransplantation

Deontological Antithesis

Justice as fairness requires that the burdens and benefits of society be distributed equally. The means of survival ought to be shared, especially in view of sparse resources. Therefore, retransplantation should be avoided.

Consequentialist Antithesis.

Resources ought not to be squandered but distributed in ways that will have beneficial effects for the greatest possible number. Since the success rate for first transplants is generally higher than for retransplants, those who have not yet received a graft are more likely to benefit and should be selected as recipients over and above retransplant recipients.

Beyond the Limits of Self-Interest

The unusual quality of the arguments for and against retransplantation is that all four claims seem to be justified. The result is an irresolvable conflict that cannot be reduced to one of its components. Furthermore, this is a conflict with which physicians must live. The surgeon's courage to respect life and to make decisions in the face of contradictory demands from the patient and society cannot be reduced to private advantage. Commitment to the healing of individual patients in spite of the good's conflict with itself pushes the surgeon's respect for life beyond the limits of utility. Self-interest cannot explain the courage of the physician's commitment any more than it can explain the courage of organ donation or the courage to reform public policy. Furthermore, the significance of the physician's attempt to heal, the donor's (or donor's family) gift, and the political reformer's struggle is not reducible to its real or imagined origins. Significance comes from the intended effect of healing, the gift of life, and struggle to reform. Transplantation reaps some benefits that are contrary to self-interest and others that cannot be reduced to it. One such benefit is a gift bestowed by the same crises that threaten our self-identity and force us to rethink what it means to be human. Everyone can reap this benefit, for it is not restricted to those who are directly touched by transplantation. It is the gift of being pushed far enough beyond the limits of utility that we regain a sense of what one interpreter calls "the bond of life, the link between ourselves and others."

References

American Psychiatric Association (1987) *Diagnostic and Statistical Manual of Mental Disorders*, 3rd Ed. American Psychiatric Association, Washington, DC.

Andrews, B. L. (1984) *New Conceptions*. St. Martin's Press, New York, NY.

Anencephalic Organ Donation Committee of Loma Linda University Medical Center (1987) *Considerations of Anencephalic Infants as Organ Donors*, Dec. 18.

Aristotle. (1962) *Nichomachean Ethics* (Oswald, M., trans.) The Bobbs-Merrill Company, Indianapolis, IN.

Arras, J. and Shinnar, S. (1988) Anencephalic newborns as organ donors, *JAMA* **259**, 15.

The Babylonian Talmud (1989) (Cashdan, E., trans.) Soncino, London, UK.

Backlund, E. O., Granberg, P. O., Hamberger, B., Veith, F. J., Fein, J. M., Tendler, M. D., Veatch, R., Kleiman, M. A., and Kalkines, G. (1985) Transplantation of adrenal medullary tissue to striatum in parkinsonism, *Journal of Neurosurgery* **62**, 169–173.

Bard, A., Baker, S., and Siwolop, S. (1987) Fetal tissue transplants, *Business Week*, Dec. 7, pp. 116–123.

Basch, S. (1973) The intraphysic integration of a new organ, *Psychoanalytic Quarterly* **42**, 364–383.

Baumgartner, W. A., Traill, T. A., Cameron, D. E., Fonger, J. D., Birenbaum, I. B., and Reitz, B. A. (1989) Unique aspects of heart and lung transplantations exhibited in the "domino-donor" operation, *JAMA* **261**, 3121.

Beauchamp, T. L. and Childress, J. F. (1983) *Principles of Biomedical Ethics*, 2nd Ed. Oxford University Press, New York, NY.

Besharov, D. J. and Silver, J. D. (1987) Rationing access to advanced medical techniques, *Journal of Legal Medicine* **8**, 4.C.

Biggers, J. D. (1981) When does life begin? *The Sciences* **21**, 14.

Black, M. (1962) *Models and Metaphors*. Cornell University Press, Ithaca, NY.

Black, P. M. (1978) Brain death, *New England Journal of Medicine* **226**, 338–344, 393–401.

Black, P. M. and Zenas, N. T. (1984) Declaration of brain death in neurosurgical and neurological practice, *Neurosurgery* **15**, 170–174.

Bustillo, M., Buster, J. E., Cohen, S. W., Thorneycroft, I. H., Simon, J. A., Boyers, S. P., Marshall, J. R., Seed, R. W., Louw, J. A., and Seed, R. G. (1984) Nonsurgical ovum transfer as a treatment in infertile women, *JAMA* **251**, 1171–1173.

Cable, K. R. (1975) The tell tale heart, *Baylor Law Review* **27**, 157.

Calvin, J. (1960) *Institutes of the Christian Religion*. (McNeill, J. T., ed. and Battles, F. L., trans.) Westminster Press, Philadelphia, PA.

Caplan, A. (1987) Should fetuses or infants be utilized as organ donors? *Bioethics* **1**, 119–140.

Caplan, A. (1988) Professional Arrogance and Public Misunderstanding, *Hastings Center Report* **18**, 34–37.

Caplan, A. and Bayer, R. (1985) *Project on Organ Transplantation: Ethical, Legal and Policy Issues Pertaining to Solid Organ Procurement*. The Hastings Center, Hastings-on-Hudson, NY.

Capron, A. M. (1987) Anencephalic donors: Separate the dead from the dying, *Hastings Center Report* **17**, 5–9.

Castelnuovo-Tedesco, P. (1973) Organ transplant, body image, psychosis, *Psychoanalytic Quarterly* **42**, 349–363.

Castelnuovo-Tedesco, P. (1978) Ego vicissitudes in response to replacement or loss of body parts, *Psychoanalytic Quarterly* **47**, 381–397.

Castelnuovo-Tedesco, P. (1981) Transplantation: Psychological implications of changes in body image, in *Psychonephrology I: Psychological Factors in Hemodialysis and Transplantation* (Levy, N. B., ed.) Plenum, New York, NY.

Childress, J. (1981) Who shall live when not all can live? in *Biomedical Ethics*. (Mappes, T. and Zembaty, J., eds.) McGraw Hill, New York, NY, pp. 578–875.

Churchland, P. S. (1986) *Neurophilosophy: Toward a Unified Science of the Mind-Brain*. MIT Press, Cambridge, MA.

Clark, M., Robinson, C., and Wickelgren, I. (1988) Interchangeable parts, *Newsweek*, Sept. 12, pp. 61–63

Corea, G. (1985) *The Mother Machine*. Harper & Row, New York, NY.

Council of Europe, Parliamentary Assembly. Recommendation 1046 (1986) (1) On the use of human embryos and fetuses for diagnostic, therapeutic, scientific, industrial and commercial purposes, *Human Reproduction 1987* **2**, 67–75.

Crowne, D. P. and Marlowe, D. (1960) A new scale of social desirability independent of psychopathology, *Journal of Consulting Psychology* **24**, 349–354.

Cunningham, B. (1944) *The Morality of Organic Transplantation*. Catholic University of America Press, Washington, DC.

Curran, C. E. (1973) *Politics, Medicine and Christian Ethics*. Fortress Press, Philadelphia, PA.

Daniels, N. (1981) Health care needs and distributive justice, *Philosophy and Public Affairs* **10**, 147–179.

DeChesser, A. D. (1986) Organ donation: The supply/demand discrepancy, *Heart and Lung* **15**, 6.

Deminkov, V. P. (1962) *Experimental Transplantation of Vital Organs.* Consultants Bureau, New York, NY.

Descartes, R. (1988) *Meditations on First Philosophy: The Philosophical Writings of Descartes.* Vol II. (Cottingham, J., Stoothoff, R., and Murdoch, D., trans.) Cambridge University Press, Cambridge, UK, pp. 1–62.

Dettman, C. and Saunders, D. (1987) *The Chance of a Lifetime.* Penguin Books, Ringwood, Australia.

Edwards, R. G. (1987) Potential of research on human embryo, in *Future Aspects in Human* In Vitro *Fertilization* (Feichtinger, W. and Kemeter, P., eds.) Springer-Verlag, Berlin, Germany, pp. 245–250.

Egan, T. (1988) Rebuffed by Oregon, *The New York Times,* May 1, p. 492.

Ellul, J. (1964) *The Technological Society.* Random House, New York, NY.

Evans, R. W., Manninen, D., Overcast, T., Garrison, L., Yaga, J., Merrikin, K., and Jonsen, A., eds. (1984) *The National Heart Transplantation Study: Final Report.* Health and Population Study Center, Battelle Human Affairs Research Centers, Seattle, WA.

Feigl, H. (1960) Mind-body, *not* a pseudo-problem, in *Dimensions of Mind* (Hook, S., ed.) New York University Press, New York, NY, pp. 24–36.

Fine, A. (1986) Transplantation in the central nervous system, *Scientific American* 255, 52–58.

Fletcher, J. (1954) *Morals and Medicine.* Princeton University Press, Princeton, NJ.

Fletcher, J. (1966) *Situation Ethics.* Westminster Press, Philadelphia, PA.

Fodor, J. A. (1981) *Representations.* MIT Press, Cambridge, MA.

Fox, R. C. (1978) Organ transplantation: Sociocultural aspects, in *Encyclopedia of Bioethics,* vol. 3 (Reich, W. T., ed.) Free Press, New York, NY, pp. 1166–1169.

Freehof, S. B. (1969) *Current Reform Responsa.* Hebrew Union College Press, New York, NY.

Freehof, S. B. and Gutman, A. (1952) *Yearbook, Central Conference of American Rabbis.* Central Conference of American Rabbis, New York, NY.

Fromm, E. (1956) *The Art of Loving.* Harper and Row, New York, NY.

Fuller, E., ed. (1980) *2500 Anecdotes for All Occasions.* Avenel Books, New York, NY.

Gage, F. H. and Bjorklund, A. (1984) Intracerebral grafting of neuronal cell suspensions into the adult brain, *Central Nervous System Trauma* 1, 47–56.

Gaylin, W. (1974) Harvesting the dead, *Harper's Magazine,* Sept., pp. 23–30.

Gold, M. (1988) *And Hannah Wept: Infertility, Adoption and the Jewish Couple.* Jewish Publication Society, Philadelphia, PA.

Goldenring, J. M. (1985) The brain-life theory: Towards a consistent biological definition of humanness, *Journal of Medical Ethics* 11, 198–204.

Gorman, W. (1960) *Body Image and Image of the Brain.* Warren H. Green, St. Louis, MO.

Gotoh, M., Porter, J., Kanai, T., Monaco, A. P., and Maki, T. (1988) Multiple donor allotransplantation, *Transplantation* **45**, 1008.

Grobstein, C. (1988) *Science and the Unborn: Choosing Human Futures.* Basic Books, New York, NY.

Guth, L. (1975) History of central nervous system regeneration research, *Experimental Neurolology* **48**, 3–15.

Hanson, J. T. and Sladek, J. R. (1989) Fetal research, *Science* **246**, 775–779.

Harrison, M. R. (1986) Organ procurement for children, *Lancet* **2**, 1383–1386.

Holzgreve, W., Beller, F., Buchholz, B., Hansmann, M., and Köhler, K. (1987) Kidney transplantation from anencephalic donors, *New England Journal of Medicine* **316**, 1069–1070.

Hooker, C. A., (1981) Towards a general theory of reduction, *Dialogue* **20**, 38–59.

Horan, D. J. (1978) Euthanasia and brain death, in *Brain Death.* (Korein, J., ed.) The New York Academy of the Sciences, New York, NY, pp. 363–375.

Hull, M. R. G. (1985) Infertility, *Practitioner* **229**, 943–945.

Jacob, W. ed. (1983) *American Reform Responsa.* Central Conference of American Rabbis, New York, NY.

Jacob, W. (1987) *Contemporary Reform Response.* Central Conference of American Rabbis, New York, NY.

Janis, I. L. and Mann, L. (1977) *Decision Making: A Psychological Analysis of Conflict, Choice, and Commitment.* Free Press, New York, NY.

Janssens, L. (1983) Transplantation d'organes, *Foi et temps* **4**, 321–322.

Jonas, H. (1980) Against the stream: Comments on the definition and redefinition of death, in *Philosophical Essays: From Ancient Creed to Technological Man.* The University of Chicago Press, Chicago, IL, pp. 132–141.

Jones, H. W., Jr., Jones, G. S., Hodgen, G. D., and Rosenwaks, Z., eds. (1986) In Vitro *Fertilization-Norfolk,* Williams and Wilkins, Baltimore, MD.

Jonsen, A. Selection of transplant recipients for cardiac transplantation, in *The National Heart Transplantation Study: Final Report,* Vol. IV. (Evans, R. W., Manninen, D., Overcast, T., Garrison, L., Yaga, J., Merrikin, K., and Jonsen, A., eds.) Health and Population Study Center, Battelle Human Affairs Research Centers, Seattle, WA.

Joyce, T., Corman, H., and Grossman, M. (1988) A cost effectiveness analysis of strategies to reduce infant mortality, *Medical Care* **26**, 348–360.

Kant, I. (1969) *Foundations of the Metaphysics of Morals.* (Beck, L. W., trans. and Wolff, R. P., ed.) The Bobbs-Merrill Company, New York, NY.

Kao, R. L., Christman, E. W., Luh, S. L., Krauhs, J. M., Tyers, G. F. O., and Williams, E. H. (1980) The effects of insulin and anoxia on the metabolism of isolated mature rat cardiac myocytes, *Archives of Biochemistry and Biophysics* **203**, 587.

Kaufman, H. H. and Lynn, J. (1986) Brain death, *Neurosurgery* **19**, 850–856.

Kelly, D. F. (1979) *The Emergence of Roman Catholic Medical Ethics in North America.* Edwin Mellen Press, Lewiston, NY.

Kelly, D. F. (1985) *A Theological Basis for Health Care and Health Care Ethics.* National Association of Catholic Chaplains, Milwaukee, WI.

Kelly, D. F. (1987) Artificial hearts: An ethical solution to the donor shortage? *Health Progress,* April, pp. 24–26.

Keyes, C. D. (1989) *Foundations for an Ethic of Dignity: A Study in the Degradation of the Good.* Edwin Mellen Press, Lewiston, NY.

Kolata, G. (1988) Who gets bone marrow transplants, *The New York Times,* April 20, p. 450.

Korein, J., ed. (1978) *Brain Death: Interrelated Medical and Social Issues.* The New York Academy of the Sciences, New York, NY.

Korein, J., ed. (1986) Brain states: Death, vegetation and life, in *Anesthesia and Neurosurgery,* 2nd Ed. (Cottrell, J. E. and Turndorf, H., eds.) C. V. Mosby Company, New York, NY, pp. 293–351.

Krimmell, H. T. and Foley, M. J. (1977) Abortion: An inspection into the nature of human life and potential consequences of legalizing its destruction, *University of Cincinnati Law Review* **46**, 725–732.

Lazarus, R. S. and Folkman, S. (1984) *Stress, Appraisal, and Coping.* Springer Publishing Company, New York, NY.

Leenan, H. J. J. (1986) The legal status of embryo *in vivo* and *in vitro:* Research on and medical treatment of embryos, *Law, Medicine and Health Care* **14**, 129–132.

Lenz, S. and Lauritsen, J. G. (1982) Ultrasonically guided percutaneous aspiration of human follicles under local anaethesia: A new method of collecting oocytes for in vitro fertilization, *Fertility and Sterility* **38**, 673–677.

Levin, F. (1987) *Halacha Medical Science and Technology.* Maznaim, Jerusalem.

Levinsky, R. J. (1989) Recent advances in bone marrow transplantation, *Clinical Immunology and Immunopathology* **50**, 124.

Lockwood, M., ed. (1985) *Moral Dilemmas in Modern Medicine.* Oxford University Press, New York, NY.

Lunde, D. T. (1969) Psychiatric complications of heart transplants, *American Journal of Psychiatry* **126**, 369–373.

Lutjen, P., Trounson, A., and Wood, C. (1984) The establishment and maintenance of pregnancy using *in vitro* fertilization and embryo donation in a patient with a primary ovarian failure, *Nature* **307**, 174–178.

MacLean, P. D. (1990) *The Triune Brain in Evolution: Role in Paleocerebral Functions.* Plenum Press, New York, NY.

Madrazo, I., Drucker-Colin, R., Diaz, V., Backlund, E., Granberg, P. O., and Hamberger, B. (1987) Open microsurgical autograft of adrenal medulla to the right caudate nucleus in two patients with intractable Parkinson's disease, *New England Journal of Medicine* **316**, 831–834.

Mahowald, M. B., Silver, J., and Ratcheson, R. A. (1987) The ethical options in transplanting fetal tissue, *Hastings Center Report* **17**, 9–15.

Mai, F. M. (1986) Graft and donor denial in heart transplant recipients, *American Journal of Psychiatry* **143**, 1159–1161.

Maimonides, M. (1910) *Guide for the Perplexed* (Friedländer, M., trans.) Dutton, New York, NY.

Malcom, L. A., Kawachi, I., Jackson, R., and Bonita, R. (1988) Is the pharmacological treatment of mild to moderate hypertension cost effective in stroke prevention? *New England Journal of Medicine* **101**, 167–171.

Manninen, D. L. and Evans, R. W. (1985) Public attitudes and behavior regarding organ donation, *JAMA* **253**, 3111–3115.

Marshall, T. K. (1967) Premature burial, *Medico-Legal Journal* **35**, 14–21.

Martyn, S., Wright, R., and Clark, L. (1988) Reconsidering required request, *Hastings Center Report* **18**, 27–34.

Maximus, Inc. (1988) Evaluation of Methods Used by States to Expand the Number of Organ and Tissue Donors, Final Report Executive Summary. Prepared for the Health Resources and Services Administration, under HRSA Contract #240,86*6048

Maxwell, G. (1978) Rigid designators and mind-brain identity, in *Minnesota Studies in the Philosophy of Science*, vol. IX (Savage, C. W., ed.) University of Minnesota Press, Minneapolis, MN, pp. 365–403.

McCormick, R. (1975) Transplantation of organs: A comment on Paul Ramsey, *Theological Studies* **36**, 503–509.

McCullagh, P. (1987) *The Foetus as a Transplant Donor.* John Wiley & Sons, Chichester, UK.

The Meaning of the Glorious Koran (1953) (Pickthall, M. M., trans.) The New American Library, New York, NY.

Miller, G. W. (1971) *Moral and Ethical Implications of Human Organ Transplants.* Charles C. Thomas, Springfield, IL.

Molleret, P. and Goulon, M. (1959) Le coma dépassé, *Revue Neurologique* **101**, 3–15.

Moore, F. D. (1988) Three ethical revolutions: Ancient assumptions remodeled under pressure of transplantation, *Transplantation Proceedings* **20**, 1061–1068.

Murray, T. H. (1987) Gifts of the body and the needs of strangers, *Hastings Center Report* **17**, 30–38.

Muslin, H. L. (1971) On acquiring a kidney, *American Journal of Psychiatry* **127**, 1185–1188.

Nakatani, T., Frazier, O. H., Lammemeier, D. E., Macris, M. P., and Radovancivic, B. (1989) Heterotopic heart transplantation, *Journal of Heart Transplantation* **8**, 40.

The National Organ Transplant Act of 1984. Public Law 98-507, 98 Stat. 2340.

Nelson, J. B. and Rohricht, J. S. (1984) *Human Medicine*. Augsburg Publishing House, Minneapolis, MN.

New York State Task Force on Life and the Law. (1988) *Transplantation in New York State: The Procurement and Distribution of Organs and Tissues.* New York Task Force on Life and the Law, Government of New York, p. 105.

O'Brien, B. J., Buxton, M. J., and Ferguson, B. A. (1987) Measuring the effectiveness of heart transplant programmes: Quality of life data and their relationship to survival analysis, *Journal of Chronic Diseases* **40 (supp. 1)**, 137–158.

Omnibus Reconciliation Act of 1986 (OBRA 86). Public Law 99-509.

Overcast, T. D. (1987) Introduction, in *Organ Procurement and Transplantation Manual* (Cover, B. M. and Overcast, T. D., eds.) National Health Publications, Owings Mills, MD.

People v Eulio, 472 NE2nd 286 (1984).

People v Lyons, No. 56072 (CA Sup Ct 1974).

Plato (1968) *The Republic.* (Bloom, A., trans.) Basic Books, New York, NY.

Plough, A. L. (1986) *Borrowed Time: Artificial Organs and the Politics of Extending Lives.* Temple University Press, Philadelphia, PA.

Policy Statement of the United Network for Organ Sharing (1988) *Transplantation of Foreign Nationals.* UNOS, P. O. Box 28010, Richmond, VA 23228.

Popper, K. R. and Eccles, J. C. (1977) *The Self and Its Brain.* Springer International, New York, NY.

President's Commission for the Study of Ethical Problems in Medicine and Biomedical and Behavioral Research. (1981) *Defining Death.* US Government Printing Office, Washington, DC, pp. 55–84.

Preuss, J. (1978) *Biblical and Talmudic Medicine.* (Rosner, F., trans.) Sanhedrin Press, New York, NY.

Pribram, K. H. (1986) The cognitive revolution and mind/brain issues, *American Psychologist* **41**, 507–520.

Ramsey, P. (1967) *Deeds and Rules in Christian Ethics.* Charles Scribner's Sons, New York, NY.

Ramsey, P. (1970) *The Patient as Person: Explorations in Medical Ethics.* Yale University Press, New Haven, CT.

Refuah L' or Hahalacha, Vol. 2. (1983) Hamesora, Jerusalem.

Report of the ad hoc committee of the Harvard medical school to examine the defifinition of brain death (1984) *JAMA* **252**, 680–682.

Rolland, J. (1988) Chronic illness and the family life cycle, in *The Changing Family Life Cycle: A Framework for Family Therapy*, 2nd Ed. (Carter, B. and McGoldrick, M., eds.) Gardner Press, New York, NY, pp. 433–454.

Roodman, G. D., Vandelberg, J. L., and Kuehl, T. J. (1988) *In utero* bone marrow transplantation of fetal baboons with mismatched adult marrow, *Bone Marrow Transplantation* **3**, 141.

Rosner, F. and Bleich, J., eds. (1979) *Jewish Bioethics.* Hebrew Publishing Company, New York, NY.

Rungon, C. L., Rigg, D. L., and Grier, R. L. (1986) Allogenic tooth transplantation in the dog, *Journal of the American Veterinary Medical Association* **188**, 713.

Rutter, M. (1987) Psychosocial resilience and protective mechanisms, *American Journal of Orthopsychiatry* **57, 3**, 316–331.

Ryle, G. (1949) *The Concept of Mind.* Hutchinson & Company, London, UK.

Sachedina, A. A. (1988) Islamic views on organ transplantation, *Transplantation Proceedings* **20**, 1084–1088.

Seigel, S. (1976) The ethical dimensions of modern medicine, *United Synagogue Review*, Fall, p. 4.

Shulhan Aruch Yoreh Deah (1945) Schulsinger, New York, NY, p. 2.

Singer, P. and Kuhse, H. (1986) The ethics of embryo research. *Law, Medicine and Health Care* **14, 3–4**, 133–138.

Singer, P. and Wells, D. (1984) *The Reproductive Revolution: New Ways of Making Babies.* Oxford University Press, Oxford, UK.

Sladek, J. R., Jr. and Gash, D. M. (1984) *Neural Transplants: Development and Function.* Plenum Press, New York, NY.

Sofer, H. (1895) *Hatam Sofer Responsa Yoreh Deah.* Schlesinger, Vienna.

Sperry, R. W. (1976) Mental phenomena as causal determinants in brain function, in *Consciousness and the Brain.* (Globus, G. G., Maxwell, G., and Savodnik, I., eds.) Plenum Press, New York, NY, pp. 163–177.

Stein, K. (1987) Last rights, *Omni* Sept., p. 58.

Strachan v John F. Kennedy Memorial Hospital, A 77 (NJ Sup Ct 1986).

Strom, T. (1988) To the editor: Oregon's decision to curtail funding for organ transplantation, *New England Journal of Medicine* **319**, 1419–1420.

Strong, R., Ong, T. H., Pilloy, P., Pillay, P., Wall, D., Balderson, G., and Lynch, S. (1988) A new method of segmental orthotopic liver transplantation in children, *Surgery* **104**, 104.

Sunde, N. and Zimmer, J. (1981) Transplantation of central nervous tissue, *Acta Neurologica Scandinavica* **63**, 323–335.

Sutton, E. and Hunsicker, L. (1988) To the editor: Oregon's decision to curtail funding for organ transplantation, *New England Journal of Medicine* **319**, 1419–1420.

Tofoski, J. G., Naumov, I. M., and Saunders, D. M. (1985) *In vitro* fertilization and embryo transfer in the treatment of infertility, *Godishen Zbornik na Medicinfkiot Fakultet zo Skopje* **31**, 245–256.

Tsuji, K. T. (1988) The Buddhist view of the body and organ transplantation, *Transplantation Proceedings* **20**, 1076–1079.

Uniform Anatomical Gift Act (UAGA Model Act), National Conference of Commissioners on Uniform State Laws, 645 North Michigan Ave., Suite 510, Chicago, IL 60611, 8A U.C.A. 15 (1968).

Uniform Anatomical Gift Act with 1987 Amendments (UAGA 1987), National Conference of Commissioners on Uniform State Laws, 645 North Michigan Ave., Suite 510, Chicago, IL 60611, 8A U.C.A. 15 (1987).

Unterman, I. Y. (1955) *Shevet Miyehudah*. Rav Kook Institute, Jerusalem.

US Deparment of Health and Human Services. (1986) *Organ Transplantation: Issues and Recommendations, Report of the Task Force on Organ Transplantation*. Department of Health and Human Services, Washington, DC.

Vacanti, J. P. (1988) Beyond transplantation, *Archives of Surgery* **123**, 545.

Vacanti, J. P., Morse, M. A., Saltzmann, W. M., Domb, A. J., Perez-Atayde, A., and Langer, R. (1988) Selective cell transplantation using bioabsorbable artificial polymers as matrices, *Journal of Pediatric Surgery* **123**, 545.

Vatican Congregation for the Doctrine of the Faith. (1987) Instructions on respect for human life in its origin and the dignity of procreation, *Origins* **16**, 697, 699–711.

Vawter, D. E., Kearney, W., Gervais, K. G., Caplan, A. L., Garry, D., and Tauer, C. (1990) *The Use of Human Fetal Tissue: Scientific, Ethical, and Policy Concerns*. University of Minnesota, Minneapolis, MN.

Veatch, R. M. (1977) *Case studies in Medical Ethics*. Harvard University Press, Cambridge, MA.

Veatch, R. M. (1981) *A Theory of Medical Ethics*. Basic Books, New York, NY.

Veith, F. J. Madrazo, I., Drucker-Colin, R., Diaz, V. (1987) Brain death: A status report of medical and ethical consideration, *JAMA* **38**, 1651–1655, 1744–1748.

Waldman, H. (1988) Monoclonal antibodies for organ transplantation, *American Journal of Kidney Disease* **11**, 154.

Walker, A. E. (1985) *Cerebral Death*, 4th Ed. Urban and Schwarzenberg, Baltimore, MD.

Warnock, M. (1984) *Report of the Committee of Inquiry into Human Fertilisation and Embryology*. Her Majesty's Stationery Office, London, UK.

Warren, M. A., Maguire, D. C., and Levine, C. (1978) Can the fetus be an organ farm? *Hastings Center Report* **8**, 23–25.

Watson, J. D., Hopkins, J. H., Roberts, J. W., Steitz, J. A., and Weiner, A. M. (1987) The origins of life, in *Moleular biology of the Gene*, 4th Ed. Benjamin/Cummings Publishing Company, Menlo Park, CA.

Weil, W. B., Jr. and Benjamin, M., eds. (1987) *Ethical Issues at the Outset of Life*. Blackwell Scientific Publications, Boston, MA.

Weismeier, K., Wirth, C. J., and Milachowski, K. A. (1988) Transplantation of the meniscus, *Revue de Chirurgie Orthopedique et Reparatrice de L'Appareil Moteur* **74 (2)**, 155.

Welch, H. G. and Larson, E. B. (1988) Dealing with limited resources, *New England Journal of Medicine* **319,** 171–173.

White, R. J., Albin, M. S., Verdua, J., and Locke, G. E. (1967) The isolated monkey's brain: Operative preparation and design of support system, *Journal of Neurosurgery* **27,** 216–225.

Wilberger, J. E. (1983) Transplantation of central nervous tissue, *Neurosurgery* **13,** 90–94.

Wimsatt, W. (1976) Reductionism, levels of organization, and the mind-body problem, in *Consciousness and the Brain*. (Globus, G. G., Maxwell, G., and Savodnik, I., eds.) Plenum Press, New York, NY, pp. 205–267.

Winslade, W. J. and Ross, J. W. (1986) *Choosing Life or Death: A Guide for Patients, Families, and Professionals*. MacMillan, Inc., New York, NY.

Youngner, S. J., Allen, M., Bartlett, E. T., Cascorbi, H. F., Hau, T., Jackson, D. L., Mahowald, M. B., Martin, B. J. (1985) Psychosocial and ethical implications of organ retrieval, *New England Journal of Medicine* **313,** 321–323.

Index of Names

A

Allegheny General Hospital
Trauma Center, 48
American Academy of
Pediatrics, 49
Andrews, B. L., 101, 102
Aristotle, 7, 272
Arras, J., 236, 237

B

Baby M, 100(fn)
Bard, A., 238
Basch, S., 166
Beauchamp, T. L., 257
Besharov, D. J., 245
Black, M., 28
Brown, L., 96
Bustillo, M., 99

C

Calvin, J., 215
Caplan, A., 236, 237
Capron, A. M., 51, 236, 237
Case Western Reserve
University,
conference at, 119
Castelnuovo-Tedesco, P.,
138, 163–168, 170,
274
Chief Rabbinate, 192

Childress, J., 249, 257
Churchland, P. S., 21, 22,
28
Clark, M., 249
Congressional Office of
Technology
Assessment,
124(fn)
Council of Europe, 33(fn),
36(fn), 95(fn)
Crowne, D. P., 145
Cunningham, B., 200, 201
Curran, C. E., 213

D

Daniels, N., 249
DeChesser, A. D., 234
Deminkov, V. P., 115
Descartes, R., 20, 21

E

Eccles, J. C., 20
Egan, T., 245
Ellul, J., 216
Evans, R. W., 44

F

Feigl, H., 22
Fine, A., 122
Fletcher, J., 212, 214

Index of Subjects

A

Abortion, 208, 209, 238, 239
Allocation, 219
  ethical considerations, 256–260
  macroallocation, 259, 260
  microallocation, 257–259
  law and allocation, 241–254
  distribution network, 241(fn)
  federal and state legislation, 242, 243
  financing, 244–248
  patient selection criteria, 248–254
Alzheimer's disease, 122
Anencephaly, 15, 50–52, 56, 73–75, 206, 207, 213, 218, 236, 237
Artificial insemination, 95, 194, 197, 210, 220, 221

B

Beneficence, 15, 271–274, 277
Body parts
  interchangeability, 162–164
  symbolic of the self, 170, 171
  respect for, 55, 219
    vs mutilation, 189, 190, 195, 199–202
    use after death, 172–174, 203, 204
  and self, 161–177
  see also Monism; Dualism
Brain,
  death, *see* Death,
  determination of humanness, 25
  mind, 15–24, 269–279
  self-identity, 9, 19–29
  three stages of evolution, 25
  tissue transplants, 113–130, 207–209, 219, 220
  see also Monism; Dualism

C

Cloning, 103, 104, 106, 211